To Sara and Elena

Daniele Mundici

Logic: a Brief Course

 Springer

Daniele Mundici
Department of Mathematics and Computer Science "U. Dini"
University of Florence (Italy)

Translated by:
Krzysztof R. Apt, Centrum Wiskunde & Informatica, Amsterdam (Netherlands)

Translated from the original Italian edition:
D. Mundici: Logica: Metodo Breve, © Springer-Verlag Italia 2011

UNITEXT – La Matematica per il 3+2

ISSN print edition: 2038-5722   ISSN electronic edition: 2038-5757

ISBN 978-88-470-2360-4   ISBN 978-88-470-2361-1 (eBook)
DOI 10.1007/978-88-470-2361-1

Library of Congress Control Number: 2011942179

Springer Milan Heidelberg New York Dordrecht London

© Springer-Verlag Italia 2012

9 8 7 6 5 4 3 2 1

Cover-Design: Beatrice B, Milan

Typesetting with LaTeX: PTP-Berlin, Protago TeX-Production GmbH, Germany
(www.ptp-berlin.eu)

Springer-Verlag Italia S.r.l., Via Decembrio 28, I-20137 Milano
Springer is a part of Springer Science+Business Media (www.springer.com)

Preface

In this course proofs are given of Gödel's completeness theorem and of some of its consequences, making use of Robinson's completeness theorem and Gödel's compactness theorem for propositional logic. The reader will encounter here other key ideas of logic: a non-ambiguous syntax, the resolution method, Davis-Putnam procedure, Tarski semantics, logical equivalence and logical consequence, Herbrand models, equality axioms, Skolem normal forms, refutations viewed as graphical objects, and the construction of some nonstandard models. The mathematical prerequisites are minimal: the text is accessible to anybody who has already seen some proofs by induction.

These pages are a distillation from numerous courses of Mathematical Logic that I gave at the Department of Information Science of the University of Milan starting from 1996, and subsequently at the Department of Mathematics "Ulisse Dini" of the University of Florence. Various chapters were also tested in a course offered in the academic year 2001-2002 by Collegio Ghislieri to students of various undergraduate courses at the University of Pavia. The current text is the result of a long interaction process between the teacher and students of various cultural backgrounds. My first acknowledgements go to them.

This book can be used in a first course of Mathematical Logic for mathematicians and for computer scientists. Parts of this text can also be useful in a course of Logic for philosophers and linguists, because of numerous, never too difficult, exercises on the connection between logic and natural language. Readers wishing to continue the study of Logic will learn from this course the necessary tools to understand Gödel's incompleteness theorems, for example in the eleven chapters of the monograph of R.M. Smullyan "Gödel's Incompleteness Theorems", Oxford University Press, 1992.

I would like to thank Giulietta and Massimo Mugnai, Pierluigi Minari, Annalisa Marcja, and in particular Carlo Toffalori for their reading of the previous versions and for their suggestions.

Florence, November 2010   *Daniele Mundici*

Preface to the English edition

This is the translation of the book "Logica: Metodo Breve", published by Springer-Verlag Italia in 2011. The author is pleased to express his gratitude to Prof. Krzysztof Apt. He undertook the task of translating the book, and offered his great expertise as a computer scientist and a teacher to improve the original version in various respects.

Florence, November 2011   *Daniele Mundici*

Symbols and Expressions

The symbol $\square$ stands for the end of a proof. The symbol $\emptyset$ denotes the empty set. The set $\mathbb{N}$ of natural numbers is defined as $\mathbb{N} = \{0, 1, 2, \ldots\}$.

The adverb "not" and the conjunctions "and" and "or" play a fundamental rôle in this course and have a precise meaning that is useful to understand right from the beginning.

Essentially for reasons of convenience, "or" will always be understood in the inclusive sense, like the Latin "vel", as opposed to the construct "either ... or, but not both". This way the negation of the phrase "Luigi does not know English and cannot play piano" is "Luigi knows English or can play piano", which, as we just stipulated, leaves open the possibility that Luigi knows English and also can play piano.

Once this point is clarified the negation of "Luigi works in Florence or lives there" is "Luigi does not work in Florence and does not live there".

For simplicity the conjunction "if" will be treated in a very limited way in comparison with its multiple uses in daily language. For example, the phrase "if Luigi wins the pools, we will see him better dressed" is interpreted as "Luigi does not win the pools or we will see him better dressed". This phrase leaves open the possibility of seeing Luigi better dressed even if he does not win the pools. The negation of this phrase is "Luigi wins the pools and (=but) we will not see him better dressed". This gives us an opportunity to recall that in mathematics the conjunction "but" is flatly identified with "and".

The conjunction "if" can be found in various contexts, for instance: also if, if also, even if, only if, if only, as if. It is also used to introduce phrases that express doubt, for example "I don't know if I have studied hard enough". Often it is difficult to understand the precise meaning. In mathematics one takes care of this by assigning a precise meaning to "A only if B" intending to say "if not B, then not A", which is equivalent to "if A then B", as both are equivalent to "B or not A".

The mathematical neologism "iff" stands for "if and only if". So for example, an even number is prime iff it equals 2.

To render these parts of discourse independent of their formulations in various natural languages when we find ourselves analysing phrases, we will write systematically $\wedge$ instead of "and" and $\vee$ instead of "or". The adverb "not" is written $\neg$. Given the phrases A and B, instead of writing "if A then B" we will write $A \to B$, which, as already mentioned, stands for $\neg A \vee B$. Instead of "A iff B" we will write $A \leftrightarrow B$, which stands for $(A \to B) \wedge (B \to A)$.

How much shall we gain by writing the conjunction "and" as "$\wedge$"? As much as we have gained by writing the preposition "times" as "$\times$", when we multiply the numbers.

Contents

Part I

Propositional Logic

1

Introduction

Have a look at the following drawing:

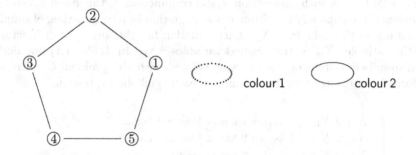

The problem $\mathcal{C}$ is to colour the vertices $1, \ldots, 5$ with one of two available colours, in such a way that vertices of the same edge have different colours. The answer is simple, but think of the same problem, denoted by $\mathcal{C}^+$, to colour a complex graph with 1000 vertices, 10000 edges and a palette of 7 colours. The search for efficient methods of solving a problem like $\mathcal{C}^+$, so either finding an explicit solution or proving that no solution exists, forms a central challenge for contemporary mathematics.

Working hypothesis. A Martian announces the solution to problems $\mathcal{C}$ and $\mathcal{C}^+$.

The unknowns, or variables, of the problem are the questions that we would ask the Martian to solve the problem ourselves. The questions have to be "binary", in the sense that the answer can be only 'yes' or 'no'. A question of the type "How many vertices have you coloured with the first colour" is not admissible. Rather, the only questions that are admissible are of the form:

Have you coloured vertex 4 with colour number 2?

For brevity we will write X_{42} instead of writing this question out in English. We will consider only the problem $\mathcal{C}$. The answers to the complete bundle of

Mundici D.: Logic: a Brief Course.
DOI 10.1007/978-88-470-2361-1_1, © Springer-Verlag Italia 2012

questions X_{ij} ($i = 1, 2, 3, 4, 5$; $j = 1, 2$) allow us to extract the solution that the Martian claims to have.

For the problem $\mathcal{C}^+$ there are 7000 unknowns X_{ij}. So each X_{ij} is a symbolic expression that expects an answer 1 (namely 'yes') or 0 (namely 'no') from the Martian. In mathematics one gives to such neither fish nor fowl expressions the name "unknowns" or "variables", to emphasise the fact that when posing the question we do not know the answer. Here there are two possible answers that depend on the solution. In traditional mathematical practice the variables and the unknowns often represent rational, real or complex numbers. Here instead, each unknown represents a *bit* (*binary digit*) that will get one of two values, 0 or 1.

The question X_{ij}, when we omit the question mark, becomes the statement "vertex i is coloured with colour number j". As such it can be negated and transformed into the statement "vertex i is not coloured with colour number j", that we will abbreviate writing $\neg X_{ij}$. These elementary statements X_{ij} and their negations $\neg X_{ij}$ are called "literals" of the problem. Operating on our 20 literals with disjunction $\vee$ and conjunction $\wedge$, the graph colouring problem $\mathcal{C}$ is completely rewritten as a system, that is, a conjunction, of simple equations in the unknowns X_{ij}. Each equation has the form of a disjunction of the variables X_{ij} or the negated variables $\neg X_{ij}$. In Table (1.1) one finds a transcribed system of equations associated with the problem $\mathcal{C}$, with an informal commentary on the meaning of each symbolic expression.

$$
\left\{
\begin{array}{ll}
X_{11} \vee X_{12} & \text{vertex 1 has at least one colour} \\
X_{21} \vee X_{22} & \text{vertex 2 has at least one colour} \\
X_{31} \vee X_{32} & \text{vertex 3 has at least one colour} \\
X_{41} \vee X_{42} & \text{vertex 4 has at least one colour} \\
X_{51} \vee X_{52} & \text{vertex 5 has at least one colour} \\
\neg X_{11} \vee \neg X_{12} & \text{vertex 1 has at most one colour} \\
\neg X_{21} \vee \neg X_{22} & \text{vertex 2 has at most one colour} \\
\neg X_{31} \vee \neg X_{32} & \text{vertex 3 has at most one colour} \\
\neg X_{41} \vee \neg X_{42} & \text{vertex 4 has at most one colour} \\
\neg X_{51} \vee \neg X_{52} & \text{vertex 5 has at most one colour} \\
\neg X_{11} \vee \neg X_{21} & \text{vertices 1 and 2 do not both have colour 1} \\
\neg X_{12} \vee \neg X_{22} & \text{vertices 1 and 2 do not both have colour 2} \\
\neg X_{21} \vee \neg X_{31} & \text{vertices 2 and 3 do not both have colour 1} \\
\neg X_{22} \vee \neg X_{32} & \text{vertices 2 and 3 do not both have colour 2} \\
\neg X_{31} \vee \neg X_{41} & \text{vertices 3 and 4 do not both have colour 1} \\
\neg X_{32} \vee \neg X_{42} & \text{vertices 3 and 4 do not both have colour 2} \\
\neg X_{41} \vee \neg X_{51} & \text{vertices 4 and 5 do not both have colour 1} \\
\neg X_{42} \vee \neg X_{52} & \text{vertices 4 and 5 do not both have colour 2} \\
\neg X_{51} \vee \neg X_{11} & \text{vertices 5 and 1 do not both have colour 1} \\
\neg X_{52} \vee \neg X_{12} & \text{vertices 5 and 1 do not both have colour 2}
\end{array}
\right.
\tag{1.1}
$$

This long rigmarole is not an example of good English prose, but it has the merit of showing that each graph colouring problem, such as $\mathcal{C}$ or $\mathcal{C}^+$, can be completely described by a few rudimentary linguistic instruments: the variables and their negations, the connective "or" and the connective "and", the latter tacitly represented by the long brace. Since the Martian understands only this language, and answers in monosyllables (bits), perhaps studying this system of equations and realising that it has no solution, he will reconsider his announcement.

In the first part of this course we will study how to decide mechanically whether a system of this type has a solution and, in case it does, how to compute at least one. Each possible solution of the system (1.1) provides sufficient information on how to colour the pentagon so that all the required conditions are satisfied.

And in fact, with a calculation given on page 29, we will verify that system (1.1) is unsatisfiable, which corresponds to the impossibility of colouring the pentagon with two colours in such a way that the vertices of the same edge have different colours. The same type of calculation will allow us to decide whether the analogous problem $\mathcal{C}^+$ has a solution. The calculation may appear useless for the trivial problem $\mathcal{C}$, but it becomes an essential instrument in solving, at least in principle, formidable problems such as $\mathcal{C}^+$. Up till now no shortcuts have been found to solve the graph colouring problems.

Appreciation of logic dawns upon us with the realisation of the fact that in formula (1.1) one does not see palettes and polygons but only a system of equations on which one works using purely symbolic manipulations, fixed and immutable, which are insensitive to the origin of the problem. So the fact that problem $\mathcal{C}$ is simple does not take away interest from its translation into system (1.1). For a problem more difficult, like $\mathcal{C}^+$, the same type of translation may provide the crucial step in obtaining a solution. As we will see, the significance of these systems goes well beyond the graph colouring problems: they are interesting objects of study, independently of the problem they represent.

Also the systems of linear equations have similar characteristics, to the point that nowadays the solvers of such systems are predominantly computer programs and not mathematicians. In logic the unknowns do not represent real numbers, but statements, which are immediate products of thought and language. Therefore to solve these systems special manipulations will be necessary, a *calculus of reasoning* (*calculus ratiocinator* in Latin).

The following table-vocabulary shows the link between problems such as $\mathcal{C}^+$ and the main concepts of mathematical logic that we will study in the next chapters.

The colouring problem	*Its formalisation*
question	variable X
answers 'no', 'yes'	truth values $0, 1$
elementary assertion, its negation	literal L
equation	clause C
system of equations for the problem	CNF formula F
a bundle of answers	assignment α
a bundle of answers solves the problem	α satisfies F
there does not exist a solution to the problem	F is unsatisfiable
there exists a solution to the problem	F is satisfiable

2

Fundamental Logical Notions

2.1 Syntax

Following the convention of the previous chapter, we use capital letters to denote *variables*. We need infinitely many variables, but by analogy with the keyboard of our computer we want to maintain our symbol apparatus, called the *alphabet*, finite. Therefore we represent officially the variables as

$$X, XI, XII, XIII, \ldots$$

To avoid lengthy sequences $III \ldots I$, we will write variables in various forms that differ from the official one, for example, using two indices X_{11}, X_{12}, X_{21}, ...:

- by a *literal* we mean a variable Y (also called a *positive literal*) or a variable preceded by the *negation* symbol $\neg Y$ (called a *negative literal*);
- by a *clause* we mean a *disjunction* of literals, that is, a sequence of the form $L_1 \vee L_2 \vee \ldots \vee L_m$;
- by a *CNF* (conjunctive normal form) formula we mean a *conjunction* of clauses, $C_1 \wedge C_2 \wedge \ldots \wedge C_k$.

Note. For each literal L, when writing $\overline{L}$ we mean the following literal, called the *opposite* of L:

$$\overline{L} = \begin{cases} \neg Y & \text{if } L \text{ coincides with the variable } Y, \\ Y & \text{if } L = \neg Y. \end{cases}$$

When writing $Var(F)$ we mean the set of the variables that occur (i.e., that appear written) in F.

Mundici D.: Logic: a Brief Course.
DOI 10.1007/978-88-470-2361-1_2, © Springer-Verlag Italia 2012

2.2 Semantics

Let F be a CNF formula. An *assignment suitable for F* is a function $\alpha : V \to \{0,1\}$, where V is a set of variables containing $Var(F)$. The set $\{0,1\}$ is called the set of *truth values*. The set V is called the *domain* of α and is denoted by $dom(\alpha)$.

We use $\{0,1\}$ to be concise and because it is easy to operate on this set by means of the functions max, min and $1-x$. But we could also use the set $\{no, yes\}$, or even more tediously, the set $\{false, true\}$.

If we think of a CNF formula as a system of equations having as many equations as clauses and whose unknowns are the variables, then an assignment is nothing else than a substitution of binary numerical values (in our case 0 and 1) for the unknowns.

Let F be a CNF formula and α an assignment suitable for F. To make precise what it means that α *satisfies* F, in symbols

$$\alpha \models F,$$

we proceed gradually, defining $\alpha \models G$ for each variable, negated variable, and for each clause of F:

(i)   if G is a variable Y, then $\alpha \models G$ means that $\alpha(Y) = 1$;

(ii) if $G = \neg Y$, then $\alpha \models G$ means that $\alpha(Y) = 0$;

(iii) if G is a clause of F, then $\alpha \models G$ means that α satisfies at least one of its literals;

(iv) finally, $\alpha \models F$ means that α satisfies each of the clauses of F.

When writing
$$\alpha \not\models F,$$
we mean that α is suitable for F and furthermore α does not satisfy F. If α is not suitable for F, it does not make sense to ask whether α satisfies F.

In (i) each variable Y takes a rôle of a receiver of a *bit*: 0 or 1. In (ii) we give meaning to the negation symbol $\neg$. In (iii) we give meaning to the disjunction $\vee$, and in (iv) we give meaning to the conjunction $\wedge$.

We say that F is *satisfiable* if some assignment α satisfies F. Otherwise we say that F is *unsatisfiable*. We say that F is a *tautology* if each assignment (suitable for F) satisfies F. A tautology is the analogue of a system of equations that holds true for any assignment of values to its unknowns.

2.3 Logical consequence and logical equivalence

One of the fundamental methods of solving a system of equations consists of transforming it into an equivalent one that is simpler. The definitions just given lend themselves well to talking about logical equivalence, by means of a notion even more important, that of logical consequence.

Logical consequence. Given two CNF formulas, F and G, we say that G is a *logical consequence* of F if for each assignment α suitable for both: if $\alpha \models F$ then $\alpha \models G$.

Logical equivalence. Two CNF formulas F and G are (logically) *equivalent*, in symbols $F \equiv G$, if they are satisfied by the same assignments suitable for both. In other words, each one is a logical consequence of the other.

Intuitively, two equivalent formulas have the same meaning. There is no risk of confusion if we say "conjunction is *commutative*", meaning that $C_1 \wedge C_2 \equiv C_2 \wedge C_1$. We will also say that "conjunction is *associative*", meaning that $C_1 \wedge (C_2 \wedge C_3) \equiv (C_1 \wedge C_2) \wedge C_3$. Furthermore, conjunction is *idempotent* in the sense that $C \wedge C \equiv C$. Analogously, disjunction is commutative, associative and idempotent.

Exercises

1. In a room there are two people (whom we will call a and b) and three musical instruments (denoted by 1, 2, 3). Using the variables

$$X_{a1}, X_{a2}, X_{a3}, X_{b1}, X_{b2}, X_{b3},$$

where X_{aj} says "a can play instrument j" and X_{bj} says "b can play instrument j" (for $j = 1, 2, 3$), write a clause, or a conjunction of clauses, to express each of the following situations:

a) the second person cannot play any instrument;
   (*Solution.* $\neg X_{b1} \wedge \neg X_{b2} \wedge \neg X_{b3}$, which is a conjunction of three clauses, each having a single literal)

b) the first person can play at least one instrument;
   (*Solution.* $X_{a1} \vee X_{a2} \vee X_{a3}$, which is a clause having three literals)

c) the first person can play exactly one instrument;
   (*Solution.* $(X_{a1} \vee X_{a2} \vee X_{a3}) \wedge (\neg X_{a1} \vee \neg X_{a2}) \wedge (\neg X_{a1} \vee \neg X_{a3}) \wedge (\neg X_{a2} \vee \neg X_{a3})$)

d) nobody (among a and b) can play the third instrument;
   (*Solution.* $\neg X_{a3} \wedge \neg X_{b3}$)

e) at most one person can play the third instrument;
   (*Solution.* $\neg X_{a3} \vee \neg X_{b3}$)

f) for each instrument there is at least one person that can play it;
   (*Solution.* $(X_{a1} \vee X_{b1}) \wedge (X_{a2} \vee X_{b2}) \wedge (X_{a3} \vee X_{b3})$)

g) somebody (always among a and b) can play the second instrument;
   (*Solution.* $X_{a2} \lor X_{b2}$)

h) each person can play at least one of the three instruments 1,2,3;

i) each person can play at most one instrument;

j) for each instrument there is at most one person who can play it;

k) each person can play exactly one of the three instruments;

l) each instrument can be played by exactly one person.

2. A committee, formed by Andrea, Beatrice and Carla, has to discuss a proposal and subsequently vote on it. Using the variables A, B, C that state respectively "Andrea, Beatrice, Carla voted for the proposal", and recalling what we said about the conjunction "if", formalise the following statements as sets of clauses:

   a) if Carla voted for the proposal, also Andrea voted for it;
      (*Solution.* $\neg C \lor A$)

   b) the vote for the proposal was unanimous;

   c) the proposal did not pass;

   d) the proposal passed, but not unanimously;

   e) only one person voted for the proposal;

   f) if Carla voted for the proposal, nobody else voted for it;

   g) only one person voted for the proposal;

   h) Carla voted differently than Andrea, but Beatrice voted like him.

3. Formalise with a set S of clauses the problem C' of bicolouring the vertices of a square, with the only condition that for any edge its two vertices have a different colour. Find an assignment that satisfies S.

4. Verify that a clause F is a tautology iff among its literals there exists a variable X and its negation $\neg X$.

   Solution. If among the literals of F there is both X and $\neg X$, then each assignment suitable for F will satisfy X or $\neg X$ and therefore will satisfy F, and hence F is a tautology. Vice versa, if for each literal L of F its opposite $\overline{L}$ does not occur in F, then the assignment defined on $Var(F)$ that satisfies each $\overline{L}$ is suitable for F and does not satisfy any literal of F, and hence F is not a tautology.

5. If the graph G has v vertices and e edges, and the palette has c colours, how many clauses will suffice to express the colourability of G?

6. Show that $\equiv$ is an equivalence relation.[1]

7. Let α be suitable for a CNF formula F. Then α satisfies F iff the restriction of α to $Var(F)$ satisfies F.

8. If B is a logical consequence of A, and C is a logical consequence of B, then C is a logical consequence of A.

9. Give an example of equivalent CNF formulas F and G such that $Var(F) \cap Var(G) = \emptyset$.

---

[1] Recall that a binary relation $\equiv$ on a set A is called an *equivalence relation* iff for all $a, b, c \in A$:
- $a \equiv a$;
- $a \equiv b$ implies $b \equiv a$;
- $a \equiv b$ and $b \equiv c$ implies $a \equiv c$.

3

The Resolution Method

3.1 Clauses and formulas as finite sets

Let F be a CNF formula. We will describe a method of transforming F into a formula equivalent to F and richer in clauses. This method then decides whether F is unsatisfiable, and whenever F is satisfiable it finds an assignment that satisfies F.

First we will simplify the notation, redefining each clause as a finite set of literals, and each CNF formula as a finite set of clauses. This is possible thanks to the commutativity, associativity and idempotence of conjunction and disjunction. In fact:

- as disjunction is associative, we are free to write $L_1 \vee \ldots \vee L_m$ without bothering to give the rules of precedence for this disjunction: such rules would not change the meaning of this clause;
- as disjunction is commutative, changing the order of its literals does not change the meaning of the clause;
- as disjunction is idempotent, we can always avoid the repetition of the same literal in a clause.

Further, as also conjunction is associative, commutative and idempotent, each CNF formula can always be viewed as the set of its clauses.

Example 3.1. Instead of $(\neg C \vee A \vee \neg C \vee B \vee \neg A \vee A \vee B)$ we will write in the new notation $\{\neg C, A, B, \neg A\}$. Instead of $(A \vee C \vee A) \wedge (A \vee \neg B) \wedge (A \vee C)$ we will write $\{\{C, A\}, \{A, \neg B\}\}$.

To facilitate the presentation of our logical calculus, the definitions of the previous chapter will now be adapted to this set-based notation.

Clause and CNF formula as sets. By a *clause* we mean a finite set of literals. By a *CNF formula* we mean a finite set of clauses.

Mundici D.: Logic: a Brief Course.
DOI 10.1007/978-88-470-2361-1_3, © Springer-Verlag Italia 2012

The semantics of CNF formulas is now naturally modified in the following way:

Semantics of CNF formulas. Let $Var(S)$ be the set of variables that occur in a set S of clauses, and let α be an assignment suitable for S, again in the sense that the domain of α contains $Var(S)$. The definition of $\alpha \models S$ (read "α *satisfies* S") proceeds as follows:

- for each variable $Y \in Var(S)$, $\alpha \models Y$ means $\alpha(Y) = 1$;

- for each negated variable $L = \neg Y$, $\alpha \models L$ means $\alpha(Y) = 0$;

- for each clause $C \in S$, where $C = \{L_1, \ldots, L_n\}$, we write $\alpha \models C$ if $\alpha \models L_j$ for some literal $L_j \in C$;

- finally, we write $\alpha \models S$ if $\alpha \models C$ for each clause $C \in S$.

Definition 3.2. Two sets of clauses S and S' are *equivalent*, in symbols, $S \equiv S'$, if they are satisfied by the same assignments suitable for both sets of clauses. We say that S' is a *(logical) consequence* of S if each assignment α suitable for S and S' that satisfies S also satisfies S'.

The introduction of the zero symbol in the Hindu-Arabic numeral system, so useful to perform fast the four arithmetic operations, entailed some adjustments in their definitions. Also the set-based redefinition of clause and of CNF formula entails a semantical readjustment: we have to give meaning to the clause with no literals and to the empty set of clauses.

Empty clause and empty set of clauses. The clause with no literals is denoted by $\square$. There is no risk of confusion using the same symbol to indicate the end of a proof. One stipulates that $\square$ is unsatisfiable.

One also introduces the empty set of clauses $\emptyset$, with the stipulation that each assignment satisfies $\emptyset$. In particular, the empty assignment satisfies the empty set of clauses, in symbols, $\emptyset \models \emptyset$.

Each assignment α is suitable for the empty clause and also for the empty set of clauses.

Lemma 3.3. *Let S' be the set obtained from a set of clauses S by removing a tautology. Then $S \equiv S'$.*

Proof. If α satisfies S, because $S' \subseteq S$, α automatically satisfies S'. Vice versa, suppose that α satisfies S' and is suitable for S. Since S is obtained from S' by adding a tautology T, α satisfies T and therefore α satisfies S. $\square$

3.2 Resolution

The mechanism of transforming a CNF formula into an equivalent one, containing more clauses, is based on a single operation:

Definition 3.4. Let C_1 and C_2 be two clauses, and suppose that a literal L satisfies the conditions $L \in C_1$, $\overline{L} \in C_2$.

Then the *resolvent of C_1 and C_2 on L and $\overline{L}$* is the clause $R(C_1, C_2; L, \overline{L})$ defined by

$$R(C_1, C_2; L, \overline{L}) = (C_1 - \{L\}) \cup (C_2 - \{\overline{L}\}).$$

So we remove the element L from C_1 and the element $\overline{L}$ from C_2, and take the union of the resulting two clauses.

Example 3.5. Suppose we have two clauses

$$C_1 = \{A, \neg B, C, D, E, \neg F\} \quad \text{and} \quad C_2 = \{\neg A, D, \neg E, G, H, Z, T\}.$$

Then the resolvent of C_1 and C_2 on A and $\neg A$ is the clause

$$(C_1 - \{A\}) \cup (C_2 - \{\neg A\}) = \{\neg B, C, D, E, \neg F, \neg E, G, H, Z, T\}.$$

Lemma 3.6 (Correctness of Resolution). *Let C_1 and C_2 be two clauses, $L \in C_1$ and $\overline{L} \in C_2$ two literals, and $D = R(C_1, C_2; L, \overline{L})$ their resolvent. Then D is a logical consequence of the conjunction $\{C_1, C_2\}$ of C_1 and C_2.*

Proof. As D does not have new variables, each assignment α suitable for $\{C_1, C_2\}$ is automatically suitable for D. Assuming $\alpha \models C_1$ and $\alpha \models C_2$ we have to show $\alpha \models D$. By assumption, $\alpha \models M$ for some $M \in C_1$, and $\alpha \models N$ for some $N \in C_2$. It is impossible that $M = L$ and $N = \overline{L}$, because then α would satisfy both the literal L and its opposite. Hence we will find at least one among M and N again as an element of D. We conclude that α satisfies that literal of D, whence $\alpha \models D$, as desired. □

We have immediately the following:

Corollary 3.7. *Given the previous assumptions, $\{C_1, C_2\} \equiv \{C_1, C_2, D\}$.*

Suppose that we start from a set S of clauses and discover that the empty clause is the resolvent of two clauses $C_1, C_2 \in S$. Then by Corollary 3.7 we have $\{C_1, C_2\} \equiv \{C_1, C_2, \square\}$. Since $\{C_1, C_2, \square\}$ is unsatisfiable, also $\{C_1, C_2\}$ is unsatisfiable. Therefore, given that $S \supseteq \{C_1, C_2\}$, also S is unsatisfiable.

Also if we do not have so much luck immediately, it could happen that after having added to S all the resolvents of the first generation, the empty clause now appears as the resolvent of two clauses from the new set S'. Consequently, the set

$$S' \cup \{ \text{ the resolvents of all the clauses in } S' \}$$

is unsatisfiable, and therefore S itself is unsatisfiable.

Of course, it could happen that the empty clause does not appear among the resolvents of the second generation, but appears among those of the third one, and so on. Lemma 3.6 tells us that, if after a certain number of steps the empty clause appears, then we have a *sufficient* condition to affirm the unsatisfiability of S. We will soon see in Theorem 4.1 that this is also a *necessary* condition.

3.3 Davis-Putnam procedure (DPP)

This procedure, or algorithm, takes as input a finite set S of clauses and produces as output the clause $\square$ (demonstrating this way that S is unsatisfiable) or else the empty set of clauses $\emptyset$; in the latter case the procedure furnishes an assignment satisfying S.

Let us see how DPP operates on an example. Suppose we have the set of clauses

$$S = \{\{A, B, \neg C\}, \{\neg A\}, \{A, B, C\}, \{A, \neg A, B, D\}, \{A, \neg B\}\}.$$

First we 'clean up' S, removing all its tautologies, according to Lemma 3.3. We are left with the set

$$S_{clean} = \{\{A, B, \neg C\}, \{\neg A\}, \{A, B, C\}, \{A, \neg B\}\}.$$

Therefore, in the preliminary stage, the original set S is simplified to the equivalent set S_{clean} without tautologies.

A step of DPP. It consists of four substeps:

1. We choose a variable that occurs in the shortest clause. In the case of a draw we use the alphabetical order. We call this selected variable the *pivot*. In our example the pivot is A.

2. We list all the A-*exempted* clauses, that is, clauses containing neither A nor $\neg A$. As there are none in our S_{clean}, the result is $\emptyset$.

3. (Among the remaining clauses) we compute all the A-*resolvents*, that is, all possible resolvents on A and $\neg A$.

4. We collect the A-exempt clauses and the A-resolvents and remove all the tautologies possibly generated in step 3. We call the resulting set S_1. In our example,
$$S_1 = \{\{B, \neg C\}, \{B, C\}, \{\neg B\}\}.$$

We have then completed the first step of DPP applied to $S_0 = S_{clean}$, using the pivot $A = P_1$. In S_1 the pivot P_1 does not occur anymore (why?) and no new variables have sneaked in. *This trivial observation guarantees that DPP terminates after a finite number of steps.*

We apply now to S_1 a second step of DPP, with the pivot $P_2 = B$: we collect the B-exempt clauses and the B-resolvents of S_1, and remove all the tautologies. We write S_2 for the resulting set and observe that the pivot P_2 does not occur in any clause of S_2. For example, in our case S_1 does not have any B-exempt clauses, there are two B-resolvents and we have $S_2 = \{\{\neg C\}, \{C\}\}$.

Proceeding this way, applying the pth step of DPP to the set of clauses S_{p-1}, we obtain the set S_p. The pivot P_p does not occur S_p. We call the variables different from P_p, that occur in S_{p-1} but not in S_p, *released*.

In our case the third step of DPP applied to the set S_2 (with the pivot $P_3 = C$) produces the set S_3 whose unique element is the empty clause, $S_3 = \{\Box\}$.

Exercises

1. Represent the following statements as clauses and apply DPP:

   If Aldo dances, then Beatrice dances. If Beatrice dances, then also Carla dances. But Carla dances only if Aldo dances. Beatrice does not dance.

2. In a room there are two people (that we will denote 1 and 2) and two chairs (also denoted 1 and 2). Using the variables

$$X_{11}, X_{12}, X_{21}, X_{22}$$

   where X_{ij} states "i sits on chair j", express in clauses the fact that each person sits on precisely one chair. Subsequently apply DPP to the set S of clauses that you wrote and find an assignment that satisfies S.

3. Formalise as a set S of clauses the problem C' of bicolouring the vertices of a square, requiring that the vertices of the same edge have a different colour. Apply DPP to S.

4. As Exercise 3, but dealing with the vertices of a triangle. Verify that DPP now yields the empty clause.

5. There are four cities, 1,2,3,4. The roads that connect 1,2,3 form an equilateral triangle with each edge 100 kilometers long. The same holds for the roads that connect 2,3,4. There are no other roads. Write the clauses for a travel plan of a salesman of carpets who starts from city 1 and visits the other three cities in three steps of hundred kilometers. Use nine variables X_{it} that state "I am in city i at the end of stage t", where $i = 2, 3, 4$ and $t = 1, 2, 3$. Apply DPP to the set S of clauses thus obtained.

6. How could one have stated Lemma 3.3 without having introduced the empty set of clauses? And how could we have stated Lemma 3.6 without having introduced the empty clause?

7. Given the variables $X_1, \ldots, X_n$, how many clauses C can we write such that $Var(C) \subseteq \{X_1, \ldots, X_n\}$?

8. Apply DPP to the set of clauses of the final example of Exercise 1 on page 10, and obtain the empty clause.

9. What could have happened in DPP if the initial set were not preventively cleaned of the tautologies? And if at the end of each step of DPP we had not removed the tautologies?

4

Robinson's Completeness Theorem

4.1 Statement and proof

As we have seen, after a number of steps t^* not exceeding the number of variables in S_0, DPP terminates producing as output a set S_{t^*} of clauses without variables. S_{t^*} can have only one of two possible forms: $S_{t^*} = \{\Box\}$ or $S_{t^*} = \emptyset$.

In the first case, by Lemma 3.6, we conclude that S is unsatisfiable. In the second case we will construct an assignment $\alpha \models S$ retracing the steps of DPP backwards.

In this sense DPP is complete:

Theorem 4.1 (Completeness Theorem, Alan Robinson, 1965). *Let S be a finite set of clauses without any tautology that forms an input for DPP. Then after a number of steps t^* not exceeding the number v of variables in S, the set S_{t^*} consists only of the empty clause or is empty. In the first case S is unsatisfiable; in the second case S is satisfiable.*

Proof. In each step a variable (the pivot) is eliminated without introducing any new ones. This ensures the existence of the set of clauses S_{t^*} without variables, with $t^* \leq v$.

We already noticed that if S_{t^*} contains the empty clause, then S is unsatisfiable. It remains to prove that *if DPP terminates with the empty set of clauses, then S is satisfiable.*

Construction. Set $S = S_0$. Let

$$S_1, S_2, \ldots, S_{t^*-1}, S_{t^*} = \emptyset$$

be the successive sets of clauses obtained in t^* steps of DPP, with $S_{t^*-1} \neq \emptyset$. Each step t has as input S_{t-1} and pivot P_t and produces as output the set S_t. In S_t the pivot P_t does not occur anymore. Recall that the variables in $Var(S_{t-1}) - Var(S_t)$ different from the pivot are said to be *released*.

Mundici D.: Logic: a Brief Course.
DOI 10.1007/978-88-470-2361-1_4, © Springer-Verlag Italia 2012

We will construct by induction a sequence of $t^* + 1$ assignments

$$\emptyset = \alpha_{t^*} \subseteq \alpha_{t^*-1} \subseteq \alpha_{t^*-2} \subseteq \ldots \subseteq \alpha_2 \subseteq \alpha_1 \subseteq \alpha_0 \qquad (4.1)$$

such that each α_t satisfies S_t, $dom(\alpha_t) = Var(S_t)$, and α_{t-1} extends α_t.

The *induction base* is the trivial observation that S_{t^*} is satisfied by the empty assignment (denoted here for consistency by α_{t^*}).

For the *induction step* we have as the induction hypothesis that the set of clauses S_t is satisfied by an assignment α_t. We will show that S_{t-1} is satisfied by an appropriate extension α_{t-1} of α_t. To begin, let ω be the extension of α_t that assigns value 0 to all variables released in step t. Let ω^- and ω^+ be the two possible extensions of ω on the pivot P_t: in other words, ω^+ assigns 1 to the pivot, while ω^- assigns to it 0. Both ω^+ and ω^- are suitable for S_{t-1}.

Claim. At least one among ω^+ and ω^- satisfies S_{t-1}.

Suppose by contradiction that the claim does not hold. Then there exist in S_{t-1} two clauses C_1 and C_2 such that

$$\omega^- \not\models C_1 \text{ and } \omega^+ \not\models C_2. \qquad (4.2)$$

Now we observe that P_t occurs in C_1, while $\neg P_t$ does not; moreover $\neg P_t$ occurs in C_2, while P_t does not. Otherwise C_1 (or C_2) would be P_t-exempt, and hence would be a clause of S_t. But then already α_t satisfies it (by the induction hypothesis) and so do its extensions ω^+ and ω^-, contrary to (4.2). Our observation is therefore confirmed.

Let D be the resolvent of C_1 and C_2 obtained by the elimination of the pivot P_t.

Case 1. D is a tautology.

Then there exists a variable X such that D contains both X and $\neg X$. (Recall Exercise 4 on page 10.) Such an X differs from the pivot P_t, and hence comes from C_1 or C_2. If X is a released variable, then both ω^+ and ω^- satisfy $\neg X$. Hence, if $\neg X \in C_1$, it follows that $\omega^- \models C_1$ contradicting (4.2). If instead $\neg X \in C_2$, it follows that $\omega^+ \models C_2$, again contradicting (4.2). Therefore X cannot be a released variable. By the induction hypothesis $X \in dom(\alpha_t)$. The assignment α_t satisfies precisely one of the literals X and $\neg X$. By construction this literal belongs to C_1 or C_2 and this way we obtain a contradiction with (4.2), recalling that both ω^+ and ω^- extend α_t. Therefore Case 1 cannot arise.

Case 2. D is not a tautology.

Then $D \in S_t$ and by the induction hypothesis $\alpha_t \models D$. It follows that $\omega \models D$. Therefore ω satisfies some literal of D, which is also satisfied by ω^+ and ω^-, and which we will find in C_1 or C_2, obtaining the usual contradiction with (4.2). So Case 2 cannot arise either and the claim is thus proved.

Defining now α_{t-1} as the first among ω^- and ω^+ that satisfies S_{t-1}, the induction step is proved. From the sequence (4.1) we extract in particular an assignment α_0 that satisfies S_0. $\qquad \square$

The essence of this proof lies in guaranteeing that the construction (4.1) never gets stuck, and hence terminates with an assignment that satisfies the initial set S. We also note that $v - t^*$ equals the number of variables that are released during the computation of DPP on S.

Example 4.2. Let $S = S_0$ be the set of clauses given by

$$S_0 = \{\{X_{11}, X_{12}\}, \{X_{21}, X_{22}\}, \{X_{31}, X_{32}\}, \{\neg X_{11}, \neg X_{12}\}, \{\neg X_{21}, \neg X_{22}\},$$

$$\{\neg X_{31}, \neg X_{32}\}, \{\neg X_{11}, \neg X_{21}\}, \{\neg X_{12}, \neg X_{22}\}, \{\neg X_{21}, \neg X_{31}\}, \{\neg X_{22}, \neg X_{32}\}\}.$$

S is already clean, that is, without any tautologies. Applying DPP we have

Step 1. pivot $P_1 = X_{11}$

$$S_1 = \{\{X_{21}, X_{22}\}, \{X_{31}, X_{32}\}, \{\neg X_{21}, \neg X_{22}\}, \{\neg X_{31}, \neg X_{32}\},$$

$$\{\neg X_{12}, \neg X_{22}\}, \{\neg X_{21}, \neg X_{31}\}, \{\neg X_{22}, \neg X_{32}\}, \{X_{12}, \neg X_{21}\}\}.$$

Step 2. pivot $P_2 = X_{12}$

$$S_2 = \{\{X_{21}, X_{22}\}, \{X_{31}, X_{32}\}, \{\neg X_{31}, \neg X_{32}\}, \{\neg X_{21}, \neg X_{31}\},$$

$$\{\neg X_{22}, \neg X_{32}\}, \{\neg X_{21}, \neg X_{22}\}\}.$$

Step 3. pivot $P_3 = X_{21}$

$$S_3 = \{\{X_{31}, X_{32}\}, \{\neg X_{31}, \neg X_{32}\}, \{\neg X_{22}, \neg X_{32}\}, \{X_{22}, \neg X_{31}\}\}.$$

Step 4. pivot $P_4 = X_{22}$

$$S_4 = \{\{X_{31}, X_{32}\}, \{\neg X_{31}, \neg X_{32}\}\}.$$

Step 5. pivot $P_5 = X_{31}$

$$S_5 = \emptyset.$$

Model-Building. Now returning to our steps and starting with the empty assignment α_5 (that automatically satisfies S_5), we will construct an assignment α_0 that satisfies S_0.

In Step 4 variable X_{32} is released. The assignment ω of this step assigns the value 0 to X_{32}. Then, to extend ω to an assignment suitable for S_4 we prepare two extensions ω^+ and ω^- on the pivot X_{31}. Theorem 4.1 guarantees that one among ω^+ and ω^- satisfies S_4. In fact, putting

$$\alpha_4(X_{31}) = 1$$

it follows that $\alpha_4 \models S_4$.

In Step 4 only the pivot is deleted. We extend α_4 to an assignment α_3 having X_{22} in its domain. Putting

$$\alpha_3(X_{22}) = 1$$

it follows that $\alpha_3 \models S_3$.

In Step 3 only the pivot is deleted. Putting $\alpha_2 \supseteq \alpha_3$ with

$$\alpha_2(X_{21}) = 0$$

it follows that $\alpha_2 \models S_2$.

In Step 2 only the pivot is deleted. Putting $\alpha_1 \supseteq \alpha_2$ with

$$\alpha_1(X_{12}) = 0$$

it follows that $\alpha_1 \models S_1$.

In Step 1 only the pivot is deleted. Putting $\alpha_0 \supseteq \alpha_1$ with

$$\alpha_0(X_{11}) = 1$$

we conclude that $\alpha_0 \models S_0$.

Following this procedure we have constructed an assignment that satisfies S. We note that S represents the problem of bicolouring three vertices $1, 2, 3$ in the graph whose two arcs connect 1 with 2 and 2 with 3. Therefore the assignment α_0 is immediately interpreted as a bicolouring of the vertices of our graph.

4.2 Refutation

Definition 4.3. Given a set of clauses S, a *refutation* of S is a finite sequence of clauses $C_1, C_2, C_3, \ldots, C_{u-1}, C_u$ in which $C_u = \square$, and each C_j belongs to S or is a resolvent of two clauses C_p and C_q with $p, q < j$.

Example 4.4. Let $S = \{\{A, B, \neg C\}, \{\neg A\}, \{A, B, C\}, \{A, \neg B\}\}$. Here is a refutation of S (with the justification of each clause, as required by the definition):

$$
\begin{aligned}
&C_1 = \{A, B, \neg C\} &&(C_1 \in S) \\
&C_2 = \{A, B, C\} &&(C_2 \in S) \\
&C_3 = \{A, B\} &&(C_3 = R(C_1, C_2; \neg C, C)) \\
&C_4 = \{A, \neg B\} &&(C_4 \in S) \\
&C_5 = \{A\} &&(C_5 = R(C_3, C_4; B, \neg B)) \\
&C_6 = \{\neg A\} &&(C_6 \in S) \\
&C_7 = \square &&(C_7 = R(C_5, C_6; A, \neg A)).
\end{aligned}
$$

Exercise 4.5. Represent this refutation as a directed graph whose vertices are the clauses and whose arcs point from each pair of generating clauses towards their resolvent.

Solution:

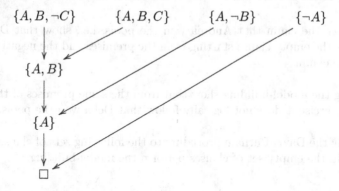

Since a refutation of S produces only logical consequences of S, the appearance of the empty clause among these consequences is a sufficient reason to conclude that no assignment satisfies S. Vice versa, the completeness theorem assures us that if S is unsatisfiable, then it has a refutation. For example, adding to S the set of all clauses produced by DPP, and calling $DPP(S)$ the resulting larger set, we obtain a refutation of S. This set, however, often contains resolvents that are useless for finding the empty clause. That is why we much appreciate *short* refutations: they retain all the power of certifying that S is unsatisfiable, and do so concisely. In addition, it takes only a few seconds to check that a simple graph like the one we have drawn above indeed represents a refutation. On the other hand, the conciseness of a good refutation has a price that we do not need to pay when we compute the resolvents with DPP: it requires *inventiveness*. The same type of inventiveness and economy is required of mathematicians in their proofs. Unfortunately, in many cases even a maximum amount of inventiveness does not succeed in reducing the length of a refutation: just as in the case of the colourability problem, there do not seem to be any shortcuts here.

Exercises

1. Write out in clauses the following premises:

    a) at least one among Andrea, Beatrice, Cesare e Delia won the pools;

    b) if Andrea won the pools, then also Cesare did;

    c) if Beatrice won the pools, then Cesare did not;

  d) if Beatrice did not win the pools, then Andrea did;

  e) if Cesare won the pools, then also Andrea or Beatrice did;

  f) if Delia won the pools, then also Cesare did;

  g) if Delia did not win the pools, Andrea or Cesare did.

Deduce the claim that Andrea won the pools, i.e., show that DPP produces the empty clause starting from the premises and the negation $\{\neg A\}$ of the claim.

2. Using the model-building show that from the same premises of the previous exercise it does not logically follow that Delia won the pools.

3. Apply the Davis-Putnam procedure to the following sets of clauses. If you obtain the empty set of clauses perform the model-building.

  a)
  $$\{\{Z, W\}, \{Z, B\}, \{\neg Z, B\}, \{W, \neg Y\},$$
  $$\{\neg Y, \neg W\}, \{\neg Y\}, \{Y, \neg B\}, \{A, \neg W, B\}, \{A, C, \neg B\},$$
  $$\{W, \neg B\}, \{\neg Y, W\}, \{Y, \neg W\}\};$$

  b)
  $$\{\{E, H\}, \{C, \neg D\}, \{Y, \neg C, \neg W\}, \{Z, \neg Y, \neg C, \neg D\},$$
  $$\{Z, \neg Y\}, \{D, \neg W\}, \{W, \neg F\}, \{F\}, \{\neg Z\}, \{\neg D, \neg C, \neg W, Z\},$$
  $$\{Y, \neg C\}, \{D, \neg Z, \neg C\}, \{\neg Y, \neg W, C\}\};$$

  c)
  $$\{\{D, \neg W\}, \{Y, \neg W, F\}, \{\neg Y, \neg W, \neg F, G\}, \{Z, \neg Y\},$$
  $$\{Y, \neg C\}, \{F, \neg G\}, \{Z, \neg C, \neg D\}, \{C, \neg D\}, \{W, \neg F\}, \{Z, C, \neg D\},$$
  $$\{\neg Z, C, \neg D\}, \{\neg Y, W, F, G\}, \{Y, \neg W, \neg F\}, \{G, \neg Z\}\};$$

  d)
  $$\{\{Y, \neg C, \neg W\}, \{Z, \neg Y, \neg C, \neg D\}, \{C, \neg D\}, \{Z, \neg Y\},$$
  $$\{D, \neg W\}, \{W, \neg F\}, \{F\}, \{\neg Z\}, \{\neg D, \neg C, \neg W, Z\}, \{Y, \neg C\},$$
  $$\{Z, C\}, \{\neg Y, \neg W\}, \{D, \neg Z\}\};$$

  e)
  $$\{\{A, \neg B\}, \{A, B, C\}, \{\neg A, \neg B, \neg C\}, \{B, \neg A\},$$
  $$\{B, \neg C\}, \{C, \neg B\}, \{B, A, \neg C\}, \{C, \neg A\}, \{A, \neg A, C\}, \{C, \neg D\}\};$$

f) $$\{\{A, \neg B, \neg C, D, \neg E\}, \{C, \neg B\}, \{D, \neg A\}, \{B\},$$

$$\{C, A, \neg E\}, \{\neg D, A, B, C, D, E\}, \{C, \neg D\}, \{\neg D, \neg A, B\},$$

$$\{E, \neg C, D\}, \{D, \neg A\}, \{C, A, \neg E\}\};$$

g) $$\{\{R, \neg S\}, \{T, \neg F\}, \{\neg Q, \neg T, \neg F, \neg G\}, \{G, \neg P\},$$

$$\{P, \neg Q\}, \{S, \neg T\}, \{Q, T, F, G\}, \{Q, \neg R\}, \{F, \neg G\}, \{P, R, S\}\}.$$

4. Given a set S of clauses, can it happen that in a refutation of S some clause of S does not appear? Can it happen that a clause is used more than once to produce several resolvents?

5. What happens if in the proof of the Completeness Theorem 4.1 we stipulate that the assignment ω assigns the value 1, instead of 0, to all the released variables?

6. Find a refutation shorter than DPP for the following set of clauses:

$$\{\{E, A\}, \{\neg B, C\}, \{\neg A, B\}, \{A, B\}\{\neg C, \neg D\}, \{\neg C, D\}, \{\neg E, \neg A, \neg C\}\},$$

and represent it as a graph, as in Exercise 4.5 on page 23.

Solution:

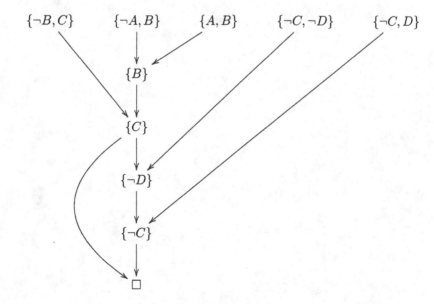

7. Find the shortest DPP refutation for the following set of clauses:

$$\{\{Q, \neg R\}, \{R, \neg S\}, \{S, \neg T\}, \{T, \neg U\}, \{U, \neg W\}, \{Q, R, S, T\}, \{T\},$$

$$\{\neg U, \neg W, \neg Z\}, \{W, \neg X\}, \{X, \neg Y\}, \{Y, \neg Z\}, \{Z, \neg Q\}, \{\neg W\}\}.$$

Represent the refutation as a graph, as in the preceding exercise.

8. Find the shortest DPP refutation for this set of clauses:

$$\{\{\neg A, B\}, \{\neg B, C\}, \{\neg C, D\}, \{\neg D, E\}, \{\neg E, F\}, \{\neg F, A\},$$

$$\{A, B, C\}, \{\neg D, \neg E, \neg F\}, \{A, C, E, F\}, \{\neg A, \neg C, \neg E, \neg F\}\}.$$

Represent the refutation as a graph.

5

Fast Classes for DPP

5.1 Krom clauses

In some cases DPP proceeds fast, also with sets K of clauses having thousands of variables, whereas other procedures (e.g., the "truth table method", in which one tries all assignments) would take geological time to decide whether K is satisfiable and to find an assignment if any.

Krom Clause. A clause having at most 2 literals is called a *Krom clause*.

In the worst cases, the first steps of DPP produce many resolvents, with a fast increment, resulting in a sort of "explosion". Yet, with Krom clauses no such explosion can occur, just because of lack of clauses. To make it precise:

Proposition 5.1. *A set S of Krom clauses with n variables is processed by DPP in at most n steps. In each step t one generates at most $2(n-t)^2 + n - t + 1$ clauses.*

Proof. In each step of DPP exactly one variable is deleted, so there are at most n steps.

After step t, there are at most $s = n - t$ variables. These variables produce exactly $2s$ literals. The possible Krom clauses are thus:

- the empty clause (with no literals);
- $2s$ clauses with 1 literal;
- $2s(2s - 1)/2$ clauses with 2 literals.

The resolvent of two Krom clauses is again a Krom clause. So after step t there are in total at most $2s^2 + s + 1 = 2(n - t)^2 + n - t + 1$ clauses. $\square$

The complete computation of DPP requires for each set S of Krom clauses an amount of space (and of computing time) moderately increasing with the number n of variables in S.

Mundici D.: Logic: a Brief Course.
DOI 10.1007/978-88-470-2361-1_5, © Springer-Verlag Italia 2012

5.2 Horn clauses

Subsumption. A clause C *subsumes* a clause G if $C \subseteq G$ and $C \neq G$. In particular, the empty clause subsumes every other clause. Also, it is easy to see that if a clause C subsumes G, then G is a logical consequence of C.

The following cleaning rule allows us to eliminate all the subsumed clauses.

Lemma 5.2. *Let S' be the set of clauses obtained from S by eliminating a clause G subsumed by another clause $C \in S$. Then $S' \equiv S$.*

Proof. If α satisfies S, then α satisfies S', because $S' \subseteq S$. Vice versa, suppose that α satisfies S' and is suitable for S. Suppose that S' is obtained from S by removing from S a clause G subsumed by some other clause $C \in S$. We note that $C \in S'$ and $\alpha \models C$, therefore α satisfies some literal $L \in C$ and hence some literal $L \in G$. Consequently, α satisfies S. $\square$

Horn Clause. A clause having at most one positive literal is called a *Horn clause*.

(Recall that positive literals were introduced in Section 2.1.) Examples of Horn clauses are the empty clause $\square$, the *unit clauses*, that is, clauses consisting of one variable $\{A\}$, $\{B\}$, $\{C\}$, ..., the completely negative clauses, for instance $\{\neg A\}$, $\{\neg A, \neg B, \neg C\}$, and the mixed clauses, for instance $\{\neg A, \neg B, C\}$, containing a single nonnegated variable and some negated variables.

Recalling Lemma 5.2 we have:

Proposition 5.3. *Let $S = S_0$ be a set of Horn clauses, without any tautologies and subsumed clauses. Suppose we delete all subsumed clauses arising during our computation of DPP. For each $t = 1, 2, \ldots,$*

(i) if S_{t-1} contains the empty clause, then S is unsatisfiable; otherwise,
(ii) if S_{t-1} does not contain a unit clause, then S is satisfiable;
(iii) if S_{t-1} contains a unit clause, then S_t is a set of Horn clauses having fewer clauses than S_{t-1}.

Proof. Case (i) is trivial. In case (ii), note that the zero assignment (assigning 0 to all variables of S_{t-1}) satisfies S_{t-1}. Arguing as in the proof of the Robinson Completeness Theorem, the zero assignment can be extended to an assignment satisfying S. In case (iii), letting $\{U\}$ denote the first unit clause of S_{t-1}, the variable U will also be the pivot of step t. There cannot be in S_{t-1} any clauses of the form $\{\neg A_1, \ldots, \neg A_k, U\}$, because they are subsumed by $\{U\}$. Thus $\{U\}$ will be used in forming each U-resolvent. Evidently, the DPP-step leading from S_{t-1} to S_t not only reduces the number of clauses but also produces shorter clauses. All clauses in S_t are Horn (why?). $\square$

Also in this case, the amount of computing time required by the complete computation of DPP is moderately increasing with the number of clauses in S.

Exercises

1. Refute the set of clauses of the formula (1.1) on page 4, that expresses the bicolourability of the vertices of the pentagon.

Solution:

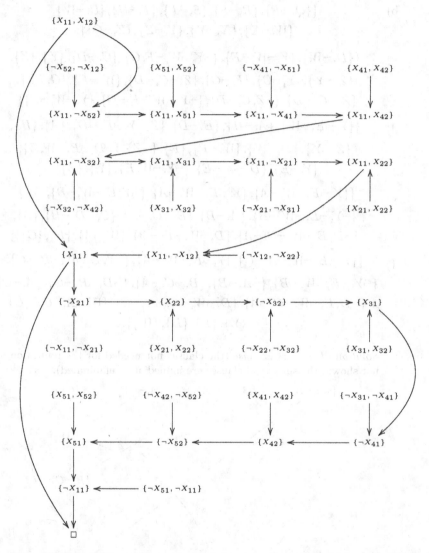

Note. Bicolourability problems are easy because they are quickly reducible to satisfiability problems for Krom clauses. One can avoid such reductions, using a combinatorial lemma to the effect that a graph G is bicolourable iff it does not have cycles with an odd number of vertices. Then, however, one will have to describe a just as fast procedure that checks that G does not have such cycles.

2. Apply DPP to the following sets of clauses. If you obtain the empty set of clauses perform the model-building.

a) $\{\{Q, \neg R\}, \{R, \neg S\}, \{S, \neg T\}, \{T, \neg U\}, \{U, \neg W\},$
$\{W, \neg X\}, \{X, \neg Y\}, \{Y, \neg Z\}, \{Z, \neg Q\}, \{T\}, \{\neg W\}\};$

b) $\{\{Q, \neg R\}, \{R, \neg S\}, \{S, \neg T\}, \{T, \neg U\}, \{U, \neg W\},$
$\{W, \neg X\}, \{X, \neg Y\}, \{Y, \neg Z\}, \{Z, \neg Q\}\};$

c) $\{\{D, \neg W\}, \{Y, \neg W, \neg F\}, \{\neg Y, \neg W, \neg F, G\}, \{C, \neg D\}, \{Y\}, \{Z\},$
$\{Z, \neg Y\}, \{Y, \neg C\}, \{F, \neg G\}, \{Z, \neg C, \neg D\}, \{W, \neg F\}, \{G, \neg Z\},$
$\{Z, \neg C, \neg D\}, \{\neg Z, C, \neg D\}, \{\neg Y, \neg W, \neg F, \neg G\}, \{Y, \neg W, \neg F\}\};$

d) $\{\{Y, \neg E, \neg W\}, \{A, \neg H\}, \{E, \neg D\}, \{Z, \neg Y, \neg E, \neg D\}, \{A\}, \{H\},$
$\{Z, \neg Y\}, \{D, \neg W\}, \{W, \neg F\}, \{F\}, \{\neg Z\}, \{\neg D, \neg E, \neg W, Z\},$
$\{Y, \neg E\}, \{D, \neg Z, \neg E\}, \{\neg Y, \neg W, E\}, \{Y\}, \{Z\}\};$

e) $\{\{Y, \neg E, \neg W, \neg A\}, \{Y, \neg E, \neg W, \neg A\}, \{\neg Y, E, \neg W, \neg B\}, \{W\},$
$\{\neg Y, \neg E, \neg W, \neg B\}, \{A, \neg B\}, \{B, \neg C, \neg A\}, \{C, \neg D, \neg B\}, \{A\},$
$\{\neg D, E, \neg W, \neg C, \neg A\}, \{D, \neg W, \neg E, \neg B\}, \{W, \neg A\}, \{B\}, \{C\}\};$

f) $\{\{Y, \neg E, \neg W, \neg A, \neg C\}, \{Y, \neg E, \neg W, \neg A\}, \{\neg Y, E, \neg W, \neg B, \neg C\},$
$\{\neg Y, \neg E, \neg W, \neg B\}, \{\neg A, \neg B\}, \{B, \neg C, \neg A\}, \{\neg D, \neg B, \neg C\}, \{A, \neg C\},$
$\{\neg D, E, \neg W, \neg C, \neg A\}, \{D, \neg W, \neg E, \neg B, \neg C\}, \{W, \neg A, \neg C\}, \{C\},$
$\{Y\}, \{E\}, \{D\}, \{W\}\}.$

Solution of the last exercise (the clauses not needed for the refutation are not shown; the subsumed clauses are immediately eliminated):

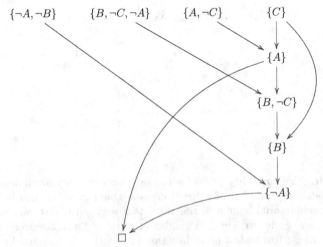

6

Gödel's Compactness Theorem

6.1 Preparatory material

So far we have considered only finite sets of clauses. But as we will see in the second part of this course, infinite sets play an important rôle. Therefore we extend the notion of satisfiability as follows:

Satisfaction of an infinite set of clauses. Let S be a (finite or infinite) set of clauses. Let $Var(S)$ denote the set of variables that occur in the clauses of S. Then an assignment α is *suitable for* S if the domain of α contains $Var(S)$. We say that α *satisfies* S, and write

$$\alpha \models S,$$

if it satisfies each clause of S; S is *unsatisfiable* if no assignment satisfies it.

We also say that a set S of clauses is *finitely satisfiable* (in short f.s.) if each *finite* subset Q of S is satisfiable. In this case the assignment $\alpha = \alpha_Q$ that satisfies Q depends on Q, and at first sight nothing suggests that by varying α_Q one can obtain an assignment α^* that simultaneously satisfies all clauses of S. Vice versa, as an assignment that satisfies S also satisfies each finite subset of it, the satisfiability of S seems to be a stronger condition than finite satisfiability.

Surprisingly, Theorem 6.2 will show that both conditions are equivalent. Before moving on to its proof here is a simple test to check our understanding of the notion of finite satisfiability:

Lemma 6.1. *Let S be a finitely satisfiable set of clauses and X a variable. Let $S' = S \cup \{\{X\}\}$ be the set obtained by adding to S the unit clause $\{X\}$. Let $S'' = S \cup \{\{\neg X\}\}$. Then at least one of S' or S'' is finitely satisfiable.*

Proof. Suppose by contradiction that neither S' nor S'' is finitely satisfiable. As $S \cup \{\{X\}\}$ is not f.s., a finite subset $\{A_1, \ldots, A_p\}$ of it is unsatisfiable, and therefore

$$\text{no assignment satisfies } \{A_1, \ldots, A_p\}. \tag{6.1}$$

Mundici D.: Logic: a Brief Course.
DOI 10.1007/978-88-470-2361-1_6, © Springer-Verlag Italia 2012

As S is f.s., it follows that the unit clause $\{X\}$ has to be one of the A_i clauses, say A_p. Therefore $\{A_1, \ldots, A_{p-1}\} \subseteq S$ is satisfiable. Analogously there exists an unsatisfiable finite subset $\{B_1, \ldots, B_q\} \subseteq S \cup \{\{\neg X\}\}$ and we can write

$$\text{no assignment satisfies } \{B_1, \ldots, B_q\}. \tag{6.2}$$

It follows that $\{\neg X\}$ has to coincide with one of the B_j clauses, say B_q. Therefore $\{B_1, \ldots, B_{q-1}\} \subseteq S$ is satisfiable. As $\{A_1, \ldots, A_{p-1}, B_1, \ldots, B_{q-1}\}$ is a finite subset of S, by assumption it is satisfiable. Let α be an assignment such that

$$\alpha \models \{A_1, \ldots, A_{p-1}, B_1, \ldots B_{q-1}\}. \tag{6.3}$$

Assuming α is also suitable for $\{X\}$ (if not, we could extend it to X in an arbitrary way), there are two cases.

Case 1. $\alpha \models X$. But then α contradicts (6.1).

Case 2. $\alpha \models \neg X$. But then α contradicts (6.2).

The contradiction we found in both cases makes the initial assumption untenable, and therefore either S' or S'' is finitely satisfiable, as we wished to prove. $\qquad\square$

6.2 Statement and proof

Theorem 6.2 (Compactness Theorem of Gödel, 1930). *If S is a countable unsatisfiable set of clauses, then some finite subset of S is unsatisfiable. As already noted, the converse implication is trivial.*

Proof. We will equivalently prove that if S is finitely satisfiable then it is satisfiable. Let $Var(S) = \{X_1, X_2, X_3, \ldots\}$ be the set of all variables of S. We construct an increasing sequence $S_0 \subseteq S_1 \subseteq S_2 \ldots$ of sets of clauses as follows:

$$S_0 = S$$

$$S_{n+1} = \begin{cases} S_n \cup \{\{X_{n+1}\}\}, & \text{if } S_n \cup \{\{X_{n+1}\}\} \text{ is f.s.} \\ S_n \cup \{\{\neg X_{n+1}\}\}, & \text{if } S_n \cup \{\{X_{n+1}\}\} \text{ is not f.s.} \end{cases}$$

Then we set

$$S_* = \bigcup S_n.$$

First Claim. For each $n = 0, 1, 2, 3, \ldots$, the set S_n is f.s.

One proves this by induction. The induction base is trivial, because $S_0 = S$ is f.s. by assumption. The induction step is as follows. By the induction hypothesis each set S_n is f.s., and we have to prove that S_{n+1} is f.s. But this follows immediately from the definition of S_{n+1}, recalling Lemma 6.1.

Second Claim. Also S_* is f.s.

Indeed, each finite subset Q of S_* is included in some S_j and since by the First Claim S_j is f.s., it follows that Q is satisfiable.

Third Claim. For each variable $X_i, i = 1, 2, 3, \ldots$, exactly one among the clauses $\{X_i\}$ and $\{\neg X_i\}$ belongs to S_*.

In fact, by construction at least one of them is in S_i and therefore in S_*. But it cannot be the case that both of them are in S_*; for otherwise, S_* would contain an unsatisfiable set formed by the clauses $\{X_i\}$ and $\{\neg X_i\}$, and hence would not be f.s., which would contradict the Second Claim.

We are now able to show that S_* is satisfiable. We will construct an assignment α that satisfies S_*. To start with we set $dom(\alpha) = Var(S_*)$. Then we stipulate for each variable X_i that

$$\alpha(X_i) = 1 \text{ iff } \{X_i\} \in S_*. \tag{6.4}$$

Fourth Claim. For each clause $F \in S_*$ we have $\alpha \models F$.

Suppose by contradiction that $\alpha \not\models F$. Let $L_1, \ldots, L_u$ be the literals of F. Therefore for all $i = 1, \ldots, u$, $\alpha \not\models L_i$, and hence α satisfies the opposite of L_i, i.e., $\alpha \models \overline{L_i}$. Combining the Third Claim with the definition (6.4) it follows that $\{\overline{L_i}\} \in S_*$. As by the Second Claim S_* is finitely satisfiable, there exists an assignment that satisfies the set $\{F, \{\overline{L_1}\}, \ldots, \{\overline{L_u}\}\}$ of clauses, which is impossible. The Fourth Claim is thus established.

Since S is a subset of S_*, it follows from the Fourth Claim that α satisfies S. The theorem is thus established. $\qquad\square$

Exercises

1. Find an infinite set of clauses with the set of variables

$$I = \{X_1, X_2, X_3, \ldots\}$$

   that is satisfied by precisely one assignment $\alpha \colon I \to \{0, 1\}$.

2. Consider the following infinite set S of clauses:

$$\{\neg X_1, X_2\}, \{\neg X_2, X_1\}, \{\neg X_2, X_3\}, \{\neg X_3, X_2\}, \ldots.$$

   Which assignments $\alpha \colon \{X_1, X_2, X_3, \ldots\} \to \{0, 1\}$ satisfy S?

3. Suppose we are given a palette with three colours 1,2,3, and an infinite graph G, whose vertices are denoted by $1, 2, 3, \ldots$. Using the variables X_{ni} that state "vertex n has colour i", one can formalise the tricolourability of G using a set of clauses. Since G is infinite, infinitely many clauses will

be needed. Prove that G is tricolourable iff each of its finite subgraphs is tricolourable. By definition, a subgraph of G is a subset of vertices of G connected by the same edges as in G.

4. Can it happen in Lemma 6.1 that both S' and S'' are finitely satisfiable?

5. What happens in Theorem 6.2 when the set $Var(S)$ is finite?

7

Propositional Logic: Syntax

7.1 Formulas

We now study a language, known as (Boolean, or classical) *propositional logic*. While it is more extended than the language of clauses considered so far, as we will see, it is not more expressive. For this language it is still possible to define precisely the concepts of satisfiability, logical equivalence and logical consequence.

The syntax of propositional logic is more complicated than that of the logic of clauses; our point of departure is a finite set of symbols, called *alphabet*, containing three fundamental *connectives*, negation ¬, conjunction ∧, and disjunction ∨. These connectives are repeatedly applied, starting with the variables. To avoid ambiguity, as in elementary algebra, one introduces parentheses to combine expressions. So we have:

Alphabet. The set Σ consisting of the seven symbols $\{X, I, \wedge, \vee, \neg,), ($\}$ is called the *alphabet*. One calls each finite sequence of symbols of Σ a *string over* Σ.

Variables. Strings of the form $X, XI, XII, XIII, \ldots$, are called *variables*.

Formulas. One defines *formulas* inductively as follows:

- each variable is a formula;
- if F and G are formulas, then $(F \wedge G)$ is a formula;
- if F and G are formulas, then $(F \vee G)$ is a formula;
- if F is a formula, then $\neg F$ is a formula.

Mundici D.: Logic: a Brief Course.
DOI 10.1007/978-88-470-2361-1_7, © Springer-Verlag Italia 2012

Readers not familiar with inductive definitions can adopt the following, equivalent definition:

Alternative definition of a formula. A formula F is a string on the alphabet Σ for which there exists a *parsing certificate*, that is, a finite sequence $S_1, S_2, \ldots, S_u$ of strings on Σ with the property that for each $j = 1, \ldots, u$ the string S_j falls into at least one of the following cases:

- S_j is a variable;
- S_j is of the form $\neg S_i$ for some previous string S_i;
- S_j is of the form $(S_p \wedge S_q)$ with $p, q < j$;
- S_j is of the form $(S_p \vee S_q)$ with $p, q < j$;

and in addition, the final string S_u coincides with F.

7.2 Unambiguity of the syntax

There is some analogy between the definition of a formula and the definition of a refutation of a set of clauses. But here no inventiveness is needed to decide instantaneously whether a string is a formula. To demonstrate this fact we need some preparatory results:

Proposition 7.1. *Each formula F is balanced, in the sense that the number of open parentheses in F equals the number of closed parentheses.*

Proof. Let $S_1, S_2, \ldots, S_u = F$ be a parsing certificate for F. We will show by induction on $j = 1, \ldots, u$ that each string S_j is balanced. In particular F will turn out to be balanced.

Induction base. S_1 is necessarily a variable $XI \ldots I$. Hence S_1 has 0 open parentheses and 0 closed parentheses and therefore is balanced.

Induction step. Suppose that the statement is true for $S_1, S_2, \ldots, S_j$. We have to prove the statement for S_{j+1}. By definition of parsing certificate, S_{j+1} falls into at least one of the following three cases:

1. S_{j+1} is a variable.
   Then the statement is true, as already noted.
2. S_{j+1} is of the type $\neg S_t$ with $t < j + 1$. By the induction hypothesis S_t is balanced. Necessarily this will be also the case for S_{j+1}, given the fact that when passing from S_t to S_{j+1} we have not added any parentheses.
3. S_{j+1} is of the type $(S_a \vee S_b)$ or $(S_a \wedge S_b)$ with $a, b < j+1$. By the induction hypothesis both S_a and S_b are balanced. By a direct inspection also S_{j+1} is balanced. $\square$

Communications
in Computer and Information Science   **946**

Commenced Publication in 2007
Founding and Former Series Editors:
Phoebe Chen, Alfredo Cuzzocrea, Xiaoyong Du, Orhun Kara, Ting Liu,
Dominik Ślęzak, and Xiaokang Yang

Editorial Board

More information about this series at http://www.springer.com/series/7899

Liang Li · Kyoko Hasegawa
Satoshi Tanaka (Eds.)

Methods and Applications for Modeling and Simulation of Complex Systems

18th Asia Simulation Conference, AsiaSim 2018
Kyoto, Japan, October 27–29, 2018
Proceedings

Springer

Editors
Liang Li
Ritsumeikan University
Kusatsu, Shiga, Japan

Satoshi Tanaka
Ritsumeikan University
Kusatsu, Shiga, Japan

Kyoko Hasegawa
Ritsumeikan University
Kusatsu, Shiga, Japan

ISSN 1865-0929 ISSN 1865-0937 (electronic)
Communications in Computer and Information Science
ISBN 978-981-13-2852-7 ISBN 978-981-13-2853-4 (eBook)
https://doi.org/10.1007/978-981-13-2853-4

Library of Congress Control Number: 2018957409

This Springer imprint is published by the registered company Springer Nature Singapore Pte Ltd.
The registered company address is: 152 Beach Road, #21-01/04 Gateway East, Singapore 189721, Singapore

Using this proposition and calling the connectives $\wedge, \vee$ *binary*, one proves by means of the same method:

Proposition 7.2. *Suppose that the formula F has some binary connectives. Then at the left-hand side of each binary connective of F there are more open parentheses than closed ones. Moreover, there is a unique binary connective at whose left-hand side the number of open parentheses exceeds by one the number of closed parentheses.*

The following theorem shows that our syntax is unambiguous:

Theorem 7.3 (Unique reading of propositional formulas). *For each formula F exactly one of the following four cases arises:*

(i)   *The initial symbol of F is X; then F coincides with a uniquely determined variable.*

(ii) *The initial symbol of F is $\neg$; then F is of the form $\neg G$, where G is a formula.*

(iii) *The initial symbol of F is the open parenthesis and F is of the form $(A \wedge B)$ with uniquely determined formulas A and B.*

(iv) *The initial symbol of F is the open parenthesis and F is of the form $(A \vee B)$ with uniquely determined formulas A and B.*

Proof. Let $S_1, S_2, \ldots, S_u = F$ be a parsing certificate for F. The initial symbol of F has to be one of the symbols $X, \neg$ or the open parenthesis.

If the initial symbol of F is X or $\neg$, one falls respectively into the first or the second case and the claim is trivial.

If the initial symbol of F is the open parenthesis, we have to distinguish between the third and fourth case and find a unique decomposition. Suppose by contradiction that F has two different readings, say $F = (A \vee B) = (P \vee Q)$. In other words, in one parsing certificate F appears as the disjunction of the formulas A and B, while in another parsing certificate it appears as the disjunction of P and Q.

We make the working hypothesis (that soon will lead to contradiction) that A has more symbols than P. Then the $\vee$ symbol of the reading $(A \vee B)$ is to the right of the $\vee$ symbol of the reading $(P \vee Q)$. Therefore this latter $\vee$ lies in A. By Proposition 7.2 the number of open parentheses to the left of this $\vee$ in A is strictly larger than the number of closed parentheses. So in F this $\vee$ has to its left at least two more open parentheses than closed ones, because one needs to add to the parentheses of A the initial open parenthesis of F.

But at the same time the reading $(P \vee Q)$ tells us that to the left of the same $\vee$ there is one more open parenthesis than closed one, as P is balanced. This is a contradiction. So A does not have more symbols than P. By symmetry P does not have more symbols than A. So A and P have the same number of symbols and therefore the strings P and A coincide. It follows automatically that B and Q coincide.

One similarly shows that F does not admit two readings $F = (C \vee D) = (R \wedge S)$ or $F = (C \wedge D) = (R \wedge S)$. $\qquad\square$

Example 7.4. The formula $F = (\neg(XII \wedge XI) \vee XIII)$ starts with an open parenthesis "(", so we are in Case 3 or 4. Starting from the left, for each connective $\wedge$ or $\vee$ we count by how much the number of open parentheses to its left exceeds the number of closed ones. When we encounter the first binary connective for which the difference is one, we split the formula in two, the one on the left and the one on the right, eliminating the two outer parentheses of F.

In our example this connective is $\vee$, so we are in Case 4. To the left we find the formula $\neg(XII \wedge XI)$, while to the right we find the formula XIII. This last formula cannot be anymore subdivided and we are in Case 1.

On the other hand for the formula $\neg(XII \wedge XI)$ we repeat this parsing procedure. It starts with $\neg$, so we are in Case 2. We eliminate $\neg$, remaining with $(XII \wedge XI)$ to be dealt with subsequently using Case 3. At its left we remain with the variable XII, whereas at its right we have the variable XI.

The following parsing certificate records this construction:

$$XII, \; XI, \; XIII, \; (XII \wedge XI), \; \neg(XII \wedge XI), \; (\neg(XII \wedge XI) \vee XIII).$$

This certificate is represented graphically by the *parsing tree*[1] given in the following figure.

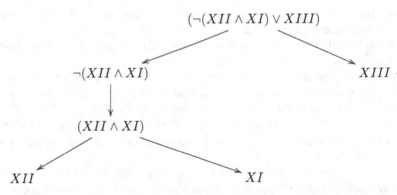

When for some string this parsing procedure gets stuck and does not succeed to reach the variables, it means that this string is not a formula. For example, the string $\neg(\neg((XI \wedge XII) \vee XIII))$ is not a formula because it does not have a binary connective to the left of which the number of open parentheses is one more than the number of closed parentheses.

---

[1] In mathematical texts, for typographical reasons, trees grow downwards.

Exercises

1. Prove Proposition 7.2.

2. Which of the following strings are formulas and which aren't? (for the latter ones the parsing procedure gets stuck):

$$(((\neg XIII \lor \neg\neg XII) \land \neg X) \lor \neg XIII), \quad (\neg XIII), \quad ((X \lor X)),$$

$$\neg(\neg(\neg X \land (XIII \lor X) \lor XII)), \quad (\neg XI \land XII),$$

$$((\neg XII \lor XII)) \lor XI) \lor (((\neg(\neg XIII) \land X) \lor XII)).$$

3. Prove by induction on the number of connectives:
   a) in no formula there occurs the sequence of symbols ();
   b) in each formula the number of open parentheses cannot be smaller than the number of $\lor$;
   c) in no formula there occurs the sequence of symbols $X\neg$;
   d) in each formula there are twice as many parentheses as binary connectives;
   e) in each formula there are fewer binary connectives than variables;
   f) if a formula does not have negation symbols and X is its only variable, then it has $4k + 1$ symbols, for some $k = 0, 1, 2, 3, \ldots$;
   g) we call a binary connective in a formula *yellow* if to its left there are precisely two more open parentheses than closed ones. Show that each formula has at most two yellow binary connectives.

4. Let p be the number of parentheses in a formula, b the number of binary connectives, and v the number of variables occurring in the formula. Prove that $p + b \geq 2(v - 1)$.

5. Confirm or refute the following statements:
   a) in no formula there occurs a binary connective to the left of which there are precisely three more open parentheses than closed ones;
   b) for each formula let v be the number of occurrences of variables in it and o and c respectively the number of open and closed parentheses. Then $3o - 2c < v$.

8

Propositional Logic: Semantics

8.1 Assignment, logical consequence, logical equivalence

Having completed the syntactic definitions we now pass on to the semantic definitions.

Assignment. By an *assignment* we mean, as always, a function α whose domain $dom(\alpha)$ is a set of variables and whose possible values are 0 and 1.

Let F be a formula defined on the alphabet Σ and let $\{X_1, \ldots, X_n\} = Var(F)$ be the set of its variables. Then the assignment α is *suitable for F* if $dom(\alpha) \supseteq \{X_1, \ldots, X_n\}$. So α assigns a truth value to each variable of F.

We will always tacitly assume that all assignments are suitable for all formulas to which they refer. And of course, when constructing assignments, we will guarantee their suitability.

Satisfiability. Given a formula F with its parsing certificate, one defines by induction $\alpha \models F$ (read: α *satisfies* F) as follows:

– if $F = Y$, where Y is a variable, then $\alpha \models Y$ means that $\alpha(Y) = 1$;

– if F is a negated formula, for example $F = \neg G$, then $\alpha \models F$ means that it is not true that $\alpha \models G$, in symbols $\alpha \not\models G$;

– if F is a conjunction, $F = (P \wedge Q)$, then $\alpha \models F$ means that $\alpha \models P$ and $\alpha \models Q$;

– if F is a disjunction, $F = (P \vee Q)$, then $\alpha \models F$ means that $\alpha \models P$ or $\alpha \models Q$.

Note. The unique reading Theorem 7.3 is crucial for the correct functioning of this definition. Without it the fact that α satisfies F could depend on the parsing certificate of F.

Mundici D.: Logic: a Brief Course.
DOI 10.1007/978-88-470-2361-1_8, © Springer-Verlag Italia 2012

Satisfiable. When there exists an assignment α such that $\alpha \models F$, then F is called *satisfiable* and otherwise it is called *unsatisfiable*.

Tautology. When $\alpha \models F$ for each assignment α (suitable for F), we say that F is a *tautology*. Clearly, F is a tautology iff $\neg F$ is unsatisfiable.

Satisfaction of an infinite set of formulas. Assume a (possibly infinite) set S of formulas. Let $Var(S)$ be the set of variables that occur in S. Then an assignment α is *suitable* for S if $dom(\alpha) \supseteq Var(S)$.

We say that α *satisfies* S, and write $\alpha \models S$, if it satisfies each formula of S. When no assignment satisfies S, we say that S is *unsatisfiable*.

Logical consequence and logical equivalence. We say that G is a *logical consequence* of F if each assignment α that satisfies F also satisfies G. It is understood that α is suitable for both formulas. Two formulas F and G are said to be *equivalent*, in symbols $F \equiv G$, if each is a logical consequence of the other.

There is an obvious relationship between logical consequence and unsatisfiability:

Lemma 8.1. *G is a logical consequence of F iff $F \wedge \neg G$ is unsatisfiable.*

This lemma is at the base of the "refutation method" for computing logical consequences, that we will study in depth later.

Exercises

In these exercises, for the sake of readability, outer parentheses of formulas will be omitted. To save other parentheses, we further assume that negation takes precedence over all binary connectives. Therefore $\neg A \vee B$ stands for $(\neg A \vee B)$ and not for $\neg(A \vee B)$.

Connectives

1. In this exercise each phrase can be written in exactly one of two ways $A \rightarrow B$ or $B \rightarrow A$. Indicate which one is the case:

   a) A is a sufficient condition for B;

   b) A is a necessary condition for B;

   c) A if B;

   d) A holds whenever B holds;

   e) A holds only when B holds;

f) A or else $\neg B$;

g) it is impossible to have A and $\neg B$;

h) A only if B;

i) A follows from B.

2. Denote by A the phrase "Arnaldo loves nature", by B the phrase "Arnaldo is a poacher", and by C the phrase "Arnaldo is silly".
   Then translate the following phrases to the symbolic language of propositional logic:

   a) if Arnaldo loves nature and is a poacher, then he is silly;

   b) Arnaldo does not love nature and is silly if he is a poacher;

   c) Arnaldo loves nature or is a poacher and is silly;

   d) if Arnaldo is a poacher or does not love nature, then he is silly.

3. Translate the following formulas to English phrases, giving to A, B, C the meaning attributed to them in Exercise 2:

   a) $(A \rightarrow B) \rightarrow \neg C$;

   b) $\neg B \wedge C$;

   c) $\neg (C \wedge \neg C)$;

   d) $(A \wedge B) \rightarrow (C \vee \neg B)$;

   e) $\neg A \vee C$;

   f) $\neg (A \vee C)$.

Assignment, logical consequence, logical equivalence

1. Given two arbitrary formulas D and E and an assignment α suitable for both of them, which of the following statements are right and which are wrong?

   a) if $\alpha \models D \wedge E$, then $\alpha \models D$ and $\alpha \models E$;

   b) if $\alpha \models D$ and $\alpha \models E$, then $\alpha \models D \wedge E$;

   c) if $\alpha \models \neg D$, then it cannot be the case that $\alpha \models D$;

   d) if it is not true that $\alpha \models D$, then $\alpha \models \neg D$;

   e) if $\alpha \models D \rightarrow E$, and moreover $\alpha \models D$, then $\alpha \models E$;

   f) if not $\alpha \models D \rightarrow E$, then $\alpha \models D$ and $\alpha \models \neg E$;

g) if $\alpha \models \neg(D \to E)$, then $\alpha \models D$ and $\alpha \models \neg E$;

h) if not $\alpha \models D \wedge E$, then $\alpha \models \neg D$ or $\alpha \models \neg E$;

i) if $\alpha \models \neg(D \wedge E)$, then $\alpha \models \neg D$ or $\alpha \models \neg E$;

j) if not $\alpha \models D \vee E$, then $\alpha \models \neg D$ and $\alpha \models \neg E$;

k) if $\alpha \models \neg(D \vee E)$, then $\alpha \models \neg D$ and $\alpha \models \neg E$.

2. For the following formulas you have at your disposal three alternatives:
   (i)   the formula is not satisfiable;
   (ii) the formula is satisfiable, but not by all assignments suitable for it;
   (iii) the formula is a tautology, that is, it is satisfied by all assignments suitable for it.

   Assign the appropriate alternative to each formula:

   a) $P \to \neg P$;

   b) $P \to Q$;

   c) $(P \to \neg P) \wedge (\neg P \to P)$;

   d) $P \to (P \wedge Q)$;

   e) $P \to (P \vee Q)$;

   f) $\neg(P \vee Q) \to (\neg Q \wedge \neg P)$;

   g) $(P \vee Q) \to (P \wedge Q)$;

   h) $(P \to Q) \to (\neg Q \to \neg P)$;

   i) $(P \to Q) \to (\neg P \to \neg Q)$;

   j) $\neg(P \to Q) \to (P \wedge \neg Q)$;

   k) $\neg(P \wedge Q) \to (\neg P \vee \neg Q)$.

3. Verify the following tautologies:
   a) $(A \to B) \to ((B \to C) \to (A \to C))$;

   (*Solution.* There are eight possible assignments for three variables; you need to verify that each of them satisfies the formula)

   b) $((A \wedge B) \to C) \to (A \to (B \to C))$;

   c) $\neg A \to (A \to B)$;

   d) $(\neg(\neg F \vee G) \vee G) \to (\neg(\neg G \vee F) \vee F)$.

4. Which of the following formulas are tautologies?

   a) $(P \to Q) \to P$;

   b) $P \to (Q \to P)$;

   c) $(P \to Q) \to (\neg P \to \neg Q)$;

   d) $\neg(\neg(\neg(\neg P \vee P) \vee P) \vee P) \vee P$;

   e) $(P \to Q) \vee (Q \to P)$;

   f) $(P \to (Q \vee R)) \to ((P \to Q) \vee (P \to R))$;

   g) $((\neg Q \to \neg R) \vee P) \to \neg((P \wedge Q) \to R)$.

5. Which of the following statements are true?

   a) if $A \vee B$ is a tautology, then at least one of A or B is a tautology;

   b) if $A \wedge B$ is a tautology, then both A and B are tautologies;

   c) for each A, it holds that A or $\neg A$ is a tautology;

   d) for each A, it holds that $A \vee \neg A$ is a tautology;

   e) for each A and B, at least one of $A \to B$ or $B \to A$ is a tautology;

   f) for each A and B, the formula $(A \to B) \vee (B \to A)$ is a tautology.

6. Verify the following logical consequences in which, by a slight abuse of notation, we write $F \models G$ to state that G is a logical consequence of F:

   a) $P \wedge (P \to Q) \models Q$ (modus ponens);

   b) $P \wedge \neg P \models Q$ (ex falso quodlibet);

   c) $\neg A \to A \models A$ (consequentia mirabilis);

   d) $\neg\neg B \to \neg A \models (A \to \neg B)$.

7. Are the following pairs of formulas equivalent?

   a) $A \to (B \to C)$, $(A \to C) \to B$;

   b) $A \to (B \to C)$, $(A \to B) \to C$;

   c) $A \to (B \vee C)$, $(A \to C) \vee (A \to B)$;

   d) $A \to (B \vee C)$, $(A \to C) \vee B$;

   e) $A \to (B \vee C)$, $\neg B \to (A \to C)$;

   f) $(A \vee B) \to C$, $(A \to C) \vee (C \to B)$.

g) $A \to \neg B$, $(B \to \neg C) \wedge C$;

h) $A \vee \neg B$, $(B \to C) \to (\neg C \to \neg B)$.

Hint. There are eight possible assignments for three variables; you need to verify that each of them satisfies the first the formula iff it satisfies the second formula.

8. Verify the following *deduction theorem*: $A \wedge B \models C$ iff $A \models B \to C$.

9. Verify the following special case of the *Craig interpolation theorem*: if $F \to G$ is a tautology and $Var(F) \cap Var(G) = \emptyset$, then F is unsatisfiable or G is a tautology.

9

Normal Forms

9.1 Some logical equivalences

We now list some logical equivalences. Their proofs follow immediately from the definitions of the previous chapter. To facilitate the reading, outer parentheses are omitted throughout:

$$F \vee G \equiv G \vee F$$
$$(F \vee G) \vee H \equiv F \vee (G \vee H)$$
$$F \vee F \equiv F$$
$$\neg\neg F \equiv F$$
$$F \vee O \equiv F \quad \text{for each unsatisfiable formula } O$$
$$F \vee \neg O \equiv \neg O \quad \text{for each unsatisfiable formula } O$$
$$\neg(\neg F \vee G) \vee G \equiv \neg(\neg G \vee F) \vee F.$$

Exercise 9.1. Prove these seven equivalences. The first three state that disjunction is commutative, associative and idempotent. Prove that the same holds for conjunction. The fourth equivalence is known as the *law of double negation*.

Exercise 9.2. Prove the following two equivalences, known as *De Morgan laws*:

$$\neg(F_1 \vee \ldots \vee F_u) \equiv \neg F_1 \wedge \ldots \wedge \neg F_u$$
$$\neg(F_1 \wedge \ldots \wedge F_u) \equiv \neg F_1 \vee \ldots \vee \neg F_u.$$

Since product distributes over sum, for each $a, \ldots, e \in \mathbb{N}$ we can write:

$$(a + b + c) \cdot (d + e) = ad + bd + cd + ae + be + ce.$$

Mundici D.: Logic: a Brief Course.
DOI 10.1007/978-88-470-2361-1_9, © Springer-Verlag Italia 2012

Analogously, we have the following *distributivity property*:

$$(F_1 \vee \ldots \vee F_u) \wedge (G_1 \vee \ldots \vee G_w) \equiv \bigvee_{i,j}(F_i \wedge G_j).$$

We also have

$$(F_1 \wedge \ldots \wedge F_u) \vee (G_1 \wedge \ldots \wedge G_w) \equiv \bigwedge_{i,j}(F_i \vee G_j).$$

Exercise 9.3. Verify these two laws of distributivity.

9.2 Propositional logic and the logic of clauses

Theorem 9.4 (CNF and DNF reduction). *For each formula F there exists an equivalent formula in CNF (understood as a conjunction of disjunctions of literals) and an equivalent formula in DNF (i.e., a disjunction of conjunctions of literals).*

Proof. We proceed by induction on the number n of the connectives in F.

Induction base. $n = 0$, that is, F is a variable X. In this case already $\{\{X\}\}$ is the desired formula in CNF equivalent to F: it is a conjunction containing a single clause, the one containing just the variable X. Analogously, $\{\{X\}\}$ is the desired formula in DNF.

Induction step. By the induction hypothesis for each formula G having $0, 1, \ldots, n$ connectives, we have equivalent formulas in CNF and DNF. We have to find equivalent formulas in CNF and DNF for each formula F having $n + 1$ connectives.

Case 1. $F = \neg G$.

Observing that G has n connectives and applying to G the induction hypothesis, we obtain $G \equiv C_1 \wedge \ldots \wedge C_p$ for appropriate clauses C_i. Using De Morgan laws and the law of double negation we can write

$$F = \neg G \equiv \neg(C_1 \wedge \ldots \wedge C_p) \equiv \neg C_1 \vee \ldots \vee \neg C_p \equiv K_1 \vee \ldots \vee K_p,$$

where each K_i is a conjunction of literals. This gives us a formula in DNF equivalent to F. Analogously one finds a formula in CNF equivalent to F, starting from a formula in DNF equivalent to G.

Case 2. $F = G \wedge H$.

Then the formulas G and H have $\leq n$ connectives and we can apply to each of them the induction hypothesis, writing, to start with, the formulas in CNF equivalent to G and to H as follows: $G \equiv C_1 \wedge \ldots \wedge C_p$ and $H \equiv D_1 \wedge \ldots \wedge D_r$. We find immediately a formula in CNF equivalent to F, writing $F \equiv G \wedge H \equiv C_1 \wedge \ldots \wedge C_p \wedge D_1 \wedge \ldots \wedge D_r$.

It is a bit more complicated to find a formula in DNF equivalent to F. By the induction hypothesis there exists a formula in DNF equivalent to G, that is, $G \equiv K_1 \vee \ldots \vee K_q$, and a formula in DNF equivalent to H, that is, $H \equiv E_1 \vee \ldots \vee E_s$. Now, using distributivity we can write

$$F \equiv G \wedge H \equiv (K_1 \vee \ldots \vee K_q) \wedge (E_1 \vee \ldots \vee E_s) \equiv \bigvee_{i,j}(K_i \wedge E_j).$$

This final formula is a formula in DNF equivalent to F.

Case 3. $F = G \vee H$.

We argue as in Case 2. From the formulas in DNF equivalent to G and to H we immediately find a formula in DNF equivalent to F. Using distributivity and starting with formulas in CNF equivalent to G and to H, we find a formula in CNF equivalent to F. ☐

One calls a set S of formulas *finitely satisfiable* if each *finite* subset of S is satisfiable. Theorem 9.4 allows us immediately to extend Theorem 6.2 to the whole propositional logic:

Theorem 9.5 (Compactness Theorem of Gödel, 1930). *Let S be a countable set of formulas. If S is unsatisfiable, then it has a finite unsatisfiable subset. As already noted, the reverse implication is trivial.*

As proved by Maltsev, the compactness theorem holds for arbitrary sets of formulas. Following Gödel, for simplicity, we proved it only for countable sets.

Boolean Algebra. A Boolean algebra is a structure $B = (B, 0, \neg, \vee)$ that satisfies the first seven equations written at the beginning of this chapter, using $=$ instead of $\equiv$ and 0 instead of O.

In each Boolean algebra B one defines the constant 1 as $\neg 0$ and the operation $\wedge$ as $x \wedge y = \neg(\neg x \vee \neg y)$.

Boolean algebras have interesting connections with various parts of mathematics: probability, topology, and set theory. Using these algebras one obtains a presentation of propositional logic in which the syntax and the logical calculus play a much less important rôle than in the presentation given on these pages.

Weakening the definition of Boolean algebra one obtains structures that are in some cases closely connected with other parts of mathematics, and at the same time correspond to logics that are interesting for some applications. For example, if we remove the equation $F \vee F = F$ from the seven equations, with the remaining six we obtain the definition of MV-algebra that corresponds to the infinite valued logic of Łukasiewicz. In this logic a phrase repeated twice gives more information than the same phrase stated only once, just as it happens when having to transmit information in a noisy environment, we repeat the words to be better understood.

Exercises

1. Prove the seven equivalences from the beginning of this chapter.

2. Find (preferably short) DNF and CNF equivalents for the following formulas, often written in an abbreviated form:

   a) $(XI \lor (X \land (X \lor (XI \land (XI \lor X)))))$;

   b) $((XI \lor XII) \land XII) \lor X$;

   c) $(\neg(X \land (\neg X \lor (XI \land \neg XI))))$;

   d) $\neg((X \land XI) \lor X)$;

   e) $(((X \lor \neg XI) \land (\neg XI \lor XII)) \land ((XI \lor XII) \lor \neg X))$;

   f) $((P \to Q) \to R) \to (P \to R)$;

   g) $(P \lor Q) \land ((Q \land (R \lor ((R \lor S) \land P))) \lor R)$;

   h) $P \to (Q \to (R \to (S \lor T)))$;

   i) $(((A \to B) \to B) \to ((B \to A) \to A)) \to (C \lor D \lor A)$;

   j) $(P \to (Q \to R)) \to ((P \land S) \to R)$;

   k) $(A \land (B \lor (\neg A \land (\neg B \lor A)))) \lor (B \land (\neg A \lor (\neg B \land A)))$;

   l) $(A \land (B \lor (\neg A \land (\neg B \lor A)))) \lor (B \land (\neg A \lor (\neg B \land (A \lor B))))$;

   m) $A \lor (B \land (C \lor (\neg A \land (\neg B \lor \neg C))))$;

   n) $(A \lor (\neg B \land (\neg C \lor (\neg A \land (B \lor C))))) \land (A \lor \neg(B \lor C)) \land (\neg A \lor \neg(B \lor \neg C))$;

   o) $D \land (A \lor (\neg B \land (\neg C \lor (\neg A \land (B \lor C)))))$;

   p) $(A \leftrightarrow B) \leftrightarrow (B \leftrightarrow \neg A)$.

3. Consider the following list of formulas, where, as always $X \to Y$ abbreviates $\neg X \lor Y$, and $X \leftrightarrow Y$ abbreviates $(X \to Y) \land (Y \to X)$:

   $X \leftrightarrow (Y \lor Z)$
   $(X \lor Y) \to W$
   $(Y \land Z) \to (X \lor \neg W)$
   _____

   $X \land (\neg Y \lor (Z \land \neg W))$

   Call "premises" the three formulas written above the line and "conclusion" the formula written under the line. Determine whether the conclusion is a logical consequence of the premises.

   Hint. For each assignment that satisfies the three premises, we have to verify that it also satisfies the conclusion. This requires sixteen checks.

The *refutation method* offers the following alternative procedure:

(i)   put each premise P_i, $i = 1, 2, 3$ in CNF and rewrite it as a finite set C_i of clauses;

(ii) do the same for the negation of the conclusion, obtaining this way a finite set N of clauses;

(iii) apply DPP to the set $C_1 \cup C_2 \cup C_3 \cup N$ of clauses.

By Lemma 8.1 the conclusion is a consequence of the premises iff we obtain the empty clause. If we do not obtain it, applying the model-building we find an assignment that satisfies the premises and the negation of the conclusion.

4. Prove Theorem 9.5.

5. Let E, T, M be three variables, where

   a) $E = $ "Martians exist";

   b) $T = $ "Alf travels in a spaceship";

   c) $M = $ "Alf meets a Martian".

   Let F be the formula $\neg E \to \neg(T \to M)$. It states

   > *If Martians do not exist, then it is not true that if Alf travels in a spaceship, Alf meets a Martian.*

   Let F' be a formula in CNF equivalent to F, written as a set of clauses in variables E, T, M. Write the following clauses:

   F'

   $\{\neg T\}$

   _____

   $\{E\}$

   Using the refutation method of Exercise 3 verify that the conclusion is a logical consequence of the premises.

   The premise F sounds plausible: if there are no Martians, how can Alf possibly meet them during his travels in a spaceship? But if we add the premise that Alf *does not* travel in a spaceship, then from this premise it logically follows that Martians *do exist*—which is surprising. Analyse this example recalling the introductory remarks made on page VII on the use of the conjunction "if" in natural language and in mathematical language.

10

Recap: Expressivity and Efficiency

Recall the example on page 3, in which we transcribed in clauses the problem of checking whether a graph G with n vertices is k-colourable. We used $n \times k$ variables with the intention that each variable X_{ij} abbreviates the phrase "vertex i has colour j". Each variable X_{ij} also represents the question "does vertex i have colour j?", and intends to contain the yes-no answer to this question.

The phrase "each vertex has at least one colour" is transformed into the conjunction of n clauses $C_1, \ldots, C_n$, where C_i states "vertex i has at least one colour", that is, $X_{i1} \vee \ldots \vee X_{ik}$. Analogously, the phrase "each vertex has at most one colour" becomes the conjunction of n formulas $K_1, \ldots, K_n$, where K_i states "vertex i has at most one colour". In detail, one forms K_i writing for each pair of different colours k' and k'' the formula $\neg X_{ik'} \vee \neg X_{ik''}$. Finally, to state "every two vertices connected by an edge have different colours" we write for each colour and for each edge a clause stating that at least one of its vertices does not have this colour.

The result of this transcription is a formula S that completely describes the initial problem. To find a colouring we treat S as a system of equations, one equation per clause, and think of the variables as unknowns to which we assign the values 1 (for "yes") or 0 (for "no"). Each assignment $\alpha \colon \{\text{variables}\} \to \{0, 1\}$ represents an attempt of colouring the graph.

As we saw in the exercises, various other combinatorial problems can be easily transcribed as satisfiability problems of a set S of clauses. The Davis-Putnam procedure solves each system S in a finite number of steps. Nevertheless, there exist examples of sets of clauses $S_1, S_2, S_3, \ldots, S_i, \ldots$ such that the length $|S_i|$ grows proportionally to i, while the number of resolvents in $DPP(S_i)$ grows exponentially in i, hence faster than any polynomial in i, which soon exceeds the memory of any computer, present or future.

Mundici D.: Logic: a Brief Course.
DOI 10.1007/978-88-470-2361-1_10, © Springer-Verlag Italia 2012

Does there exist a revolutionary procedure that, working on an arbitrary finite set S of clauses, decides its satisfiability in a number of "steps" $\sharp(S)$ much smaller than DPP, for example $\sharp(S) \leq |S|^n$, for some fixed $n \in \mathbb{N}$? This is the first problem in the famous list of seven mathematical "Millennium Prize Problems" for our millennium.[1]

---

[1] The crucial notion of "step" is taken care of by an ingenious mathematical artefact introduced by Alan Turing in 1936.

Part II

Predicate Logic

11

The Quantifiers "There Exists" and "For All"

11.1 Introduction

The fundamental notions of natural number, zero and successor are sufficiently clear to us. Addition and multiplication are then defined inductively, using zero and successor together with the equality predicate $=$, as is done on the next page of this course. More complicated arithmetic relations and operations such as $x \leq y$, "x divides y", "x is the minimum of y and z", "x is a prime number", are definable using these fundamental notions.

So for example $x \leq y$ means that there exists z such that $x + z = y$, in symbols, $\exists z \ x + z = y$. Analogously, "x is a prime number" means that $2 \leq x$ and for all $y \leq z$ such that $y \cdot z = x$ it follows that $y = 1$, in symbols,

$$s(s(0)) \leq x \ \wedge \ \forall y \forall z \ ((y \leq z \wedge y \cdot z = x) \rightarrow y = s(0)),$$

where s denotes the successor function.

In all these expressions the variables regain their familiar use, as in systems of equations; but now in addition these variables are acted upon by the quantifiers $\exists$ and $\forall$.

Consider the following *twin primes* conjecture:

> *For every x there exists a prime number $y \geq x$ such that $y + 2$ is a prime number.*

Is it true or false?

For problems such as the colourability of a graph we have used a transcription in a simple logical language, obtaining a system of equations with binary unknowns; then we have developed a logical calculus (DPP) that allows us to decide whether such a system has a solution. The calculus proceeds by obtaining various consequences of the data, or "axioms", that define the problem. The transition from the problem to its formalisation should be made with great meticulousness, because the inference mechanism knows only how

Mundici D.: Logic: a Brief Course.
DOI 10.1007/978-88-470-2361-1_11, © Springer-Verlag Italia 2012

to work on its symbols. If we omit some data or use data that do not correspond to our problem, the inference mechanism works on a different problem than the one we want to solve.

As Pythagoras thought, natural numbers are an infinity full of mystery. For their active contemplation we need a richer symbolic apparatus than that of propositional logic. The connectives $\neg$, $\vee$, $\wedge$ now connect more complex phrases having the natural subject/predicate structure. Some predicates have two or more arguments. For example, the predicate "to be smaller than" has two arguments; when we say "x is smaller than y", x is the subject and y is a complement. One also needs symbols for such "constants" as zero, that remind us of "proper names", along with symbols for the "variables". The latter are used to quantify, for example when we say "for every x there exists some y", or when we write an identity. There also appear function symbols for successor, sum and product.

Thus, following Dedekind and Peano, to address the twin primes problem we first list the properties of zero, successor, sum and product, by writing suitable axioms for $\mathbb{N}$ and its main operations:

$$
\begin{cases}
\forall x \ 0 \neq s(x) & \text{zero is not a successor of any number} \\
\forall x \forall y \ x \neq y \to s(x) \neq s(y) & \text{different numbers have different successors} \\
\forall x \ x + 0 = x & \text{zero is the neutral element of the sum} \\
\forall x \forall y \ x + s(y) = s(x + y) & \text{inductive property of the sum} \\
\forall x \ x \cdot 0 = 0 & \text{zero property of the product} \\
\forall x \forall y \ x \cdot s(y) = x \cdot y + x & \text{inductive property of the product}
\end{cases}
$$

Unlike in the case of the colouring problem, nobody tells us here: "It is enough, the axioms are sufficient for solving the problem". At one point we might even run short of ideas and not know anymore which axioms to choose.

Logic does not tell us how to find the right axioms, not even for the natural numbers. Yet, using a list of axioms $\mathcal{A}$ slightly richer than the one given above, Euclid succeeded to prove that "for every x there exists a prime y larger than x". This theorem (i) is not as evident as the axioms of $\mathcal{A}$, but is equally true, (ii) has a fundamental rôle in mathematics and, incidentally, (iii) is a necessary condition for having an infinite number of twin primes.

To prove it we use the *refutation method*, reasoning by contradiction. So we take the negation N of what we wish to prove, "there exists some x such that all y larger than x are not prime numbers". Operating on $\mathcal{A}$ and N with a variant of DPP, that we will study in the subsequent pages, we obtain the empty clause.

One can prove all theorems using this method or equivalent ones, i.e., ones that are able to prove the same theorems. So mathematics is the art of (a) finding the axioms and definitions adhering to the reality of the concepts that we are interested to study, (b) deriving from these definitions and axioms deeper and deeper results, possibly dismissing their evidence, but never their truthfulness; and if possible (c) combining these results with results obtained from other definitions and axioms, thus showing that an activity of type (a) is not gratuitous mental gymnastics.

The fact that for tasks of type (b) and (c) there do not exist more powerful methods than those already in use in the classical world, is one of the consequences of the most important theorem of this course, the completeness theorem of Gödel, which we will prove in Chapter 16.

Working hypothesis. A Martian announces that it follows from $\mathcal{A}$ that there are arbitrarily large pairs of twin primes.

However brilliant the Martian can be, on account of the completeness theorem we know that her/his solution of the twin primes conjecture will be anyway within the reach of the down-to-earth calculus presented in the subsequent chapters.

Exercises

Following Frege, to formalise the phrase "every man is mortal" we prepare two predicate symbols M_1 and M_2, where $M_1 x$ states "x is mortal" and $M_2 y$ states "y is a man". Then we write $\forall x (M_2 x \rightarrow M_1 x)$, that is, $\forall x (\neg M_2 x \vee M_1 x)$. To formalise "some man is mortal" we will write $\exists x (M_2 x \wedge \neg M_1 x)$.

1. Let Sx mean "x is a swimmer" and Ey "y is elegant". Formalise each of the following phrases:

   a) every swimmer is elegant;

   b) some swimmer is elegant;

   c) not every swimmer is elegant;

   d) some swimmer is not elegant.

2. Let Axy mean "x admires y". Transcribe the following phrases in the appropriate logical symbolism:

   a) everybody admires somebody;

   b) somebody admires everybody;

   c) somebody is admired by everybody;

   d) nobody is admired by everybody;

   e) somebody does not admire anybody.

3. Formalise the following phrases:

   a) a dog does not bite a dog;

   b) winners never quit and quitters never win;

   c) a tiger's son has stripes;

   d) Giovanna does not vote for any racist;

   e) any raptor stays up at night;

   f) no lazy person discovers new worlds;

   g) some fishermen do not boast;

   h) a healer that everybody reveres is not necessarily honest;

   i) he who sleeps doesn't catch any fish;

   j) Figaro shaves all those who do not shave themselves;

   k) every barber shaves only those who do not shave themselves.

4. Formalise:

   a) two orthogonal lines have a common point (that is, a point that belongs to both lines);

   b) if two lines are parallel, then they do not have a common point;

   c) through each point outside a line there passes a parallel to this line.

     Solutions. $\forall x \forall y (Lx \wedge Ly \wedge Oxy \rightarrow \exists z (Pz \wedge Bzx \wedge Bzy))$, $\forall x \forall y (Lx \wedge Ly \wedge Qxy \rightarrow \neg \exists z (Pz \wedge Bzx \wedge Bzy))$, $\forall x \forall y (Px \wedge Ly \wedge \neg Bxy \rightarrow \exists z (Lz \wedge Qzy \wedge Bxz))$

5. Let

  (i) Sx mean "x is a Scot";

  (ii) Cy mean "y is a type of cheese";

  (iii) Bz mean "z is a type of beer";

  (v) Lxy mean "x likes y".

Suppose then that c stands for Carlo and d for Donatella. Formalise each of the following phrases:

   a) Donatella likes all types of cheese;

   b) some Scots like all types of cheese;

   c) Donatella likes some type of cheese;

   d) every Scot likes at least one type of cheese;

   e) there is a type of cheese that all Scots like;

   f) Carlo does not like any type of cheese;

   g) every Scot dislikes any type of cheese;

   h) every Scot likes some type of cheese and some type of beer.

6. The negation of each of these phrases can be expressed by a phrase that starts with a quantifier. Find these negative forms:

   a) each supporter of Arsenal is enthusiastic;

   b) some accountant is poor;

   c) some illiterate admires all scholars;

d) every illiterate admires some scholar;

e) for every x there is some $y > x$;

f) for every x there is some y such that for every $z > y$, $f(z) > x$.

7. Note the ambiguity of the phrase "Figaro does not take seriously a person who promises quick returns." This could stand for "any person who promises quick returns to Figaro is not taken seriously by Figaro", or "not all people who promise quick returns to Figaro are taken seriously by Figaro". Formalise both phrases.

12

Syntax of Predicate Logic

12.1 Elements of the syntax

Following the same procedure as for propositional logic, we prepare now the necessary material for writing expressions upon which the logical calculus will act. In a first phase we will only work with formulas that are similar to the clauses of propositional logic. Then the calculus will be extended to all formulas.

Definition 12.1. Our *alphabet* for predicate logic is the following finite set Σ of symbols:

- connectives: $\neg, \vee, \wedge, \rightarrow$;
- quantifiers (universal and existential) $\forall, \exists$;
- variables: $x, y, z, \ldots$
- constant symbols: $a, b, c, \ldots$;
- predicate (or relation) symbols: $P, Q, R, A, B, \ldots$;
- function symbols: $f, g, h, \ldots$;
- parentheses, comma (to facilitate the reading).

Strictly speaking, the finiteness of Σ would require us to write pedantically $c, c|, c||, \ldots$ instead of $a, b, c, \ldots$. Analogously for the variables, predicates and functions. So officially

$$\Sigma = \{\neg, \vee, \wedge, \forall, \exists, x, c, P, f, |,), (, ,\}.$$

But in these pages readability is more important than strict adherence to syntactic parsimony: thus for instance, we will write x_5 (or even better, y) instead of $x|||||$, and will consider it as a unique symbol.

To avoid further pedantry, we will simply say "predicate, function, constant" instead of "predicate symbol, function symbol, constant symbol". From the context it will be clear that each predicate has a precise number of arguments and the same for functions.

Mundici D.: Logic: a Brief Course.
DOI 10.1007/978-88-470-2361-1_12, © Springer-Verlag Italia 2012

Definition 12.2. A *term* is a *string on* Σ, that is, a finite sequence of symbols of the alphabet Σ, given by the following inductive definition:

- every constant is a term;
- every variable is a term;
- if f is a function with n arguments and $t_1, \ldots, t_n$ are terms, then $f(t_1, \ldots, t_n)$ is a term.

Exercise 12.3. Give a definition of "term" as the one given earlier for propositional formulas, using an appropriate notion of "certificate". Write down a few terms and their parsing trees, in analogy to the parsing trees of propositional formulas. Formulate a "unique reading" theorem for terms, in analogy to Theorem 7.3.

Exercise 12.4. Write the parsing tree for the term

$$h(f(g(x, z, f(y, x)), g(f(c, c), h(z), y))).$$

12.2 Formalisation in clauses

Definition 12.5.

- an *atomic formula* A is a string on Σ of the form $Pt_1 \cdots t_m$, where P is a predicate with m arguments and $t_1, \ldots, t_m$ are terms;
- by a *literal* L we mean an atomic formula A or an atomic formula preceded by the negation symbol, $\neg A$;
- a *clause* is a disjunction of literals, $L_1 \vee \ldots \vee L_u$.

When a term, a literal, or a clause does not contain any variable, one says that it is *ground*.

Example 12.6. Many phrases in the exercises of the previous chapter were *universal* statements, i.e., statements starting with "each" or "for all". These statements can be easily transcribed in clauses. Take as an example the famous phrase "every man is mortal".

Following Frege, and moving away from common practice in natural language, *we prepare a variable* x, and let it range over all possible "things" or "entities" or "beings" that exist in the universe, including mythological creatures, and the ones that are born in the minds of artists and mathematicians. We also prepare a predicate (symbol) U, so that Ux abbreviates the phrase "x is a man". Analogously, we prepare a predicate M so that Mx stands for "x is a mortal". Then

> "every man is mortal" is initially transcribed as $\forall x(Ux \rightarrow Mx)$

(read: "for every being x, if x is a man, then x is mortal"). Treating "if" as we have done in propositional logic, this phrase is equivalent to "for all beings x, x is not a man or x is mortal", that is transcribed as $\forall x(\neg Ux \vee Mx)$. Now we omit the universal quantifier $\forall$. And then

| "every man is mortal" is formalised by the clause $\neg Ux \vee Mx$ |

Likewise, the phrase "all slow predators go hungry" is first transcribed as $\forall x((Px \wedge Sx) \rightarrow Hx)$, that is, $\forall x(\neg Px \vee \neg Sx \vee Hx)$. Dropping the symbol $\forall$, the phrase is formalised by the clause $\neg Px \vee \neg Sx \vee Hx$.

As long as no existential quantifiers are used, this transcription functions well also for more complicated phrases, like "every barber shaves all those who do not shave themselves". This phrase becomes the clause $\neg Bx \vee Syy \vee Sxy$ through the following transformation: $\forall x \forall y((Bx \wedge \neg Syy) \rightarrow Sxy)$ and then $\forall x \forall y(\neg Bx \vee \neg\neg Syy \vee Sxy)$.

Omitting the universal quantifier $\forall$ in all formulas in which no existential quantifiers occur is standard mathematical practice. It suffices to recall the identity $\sin^2 x + \cos^2 x = 1$.

As already done in the propositional case, to simplify the logical calculus, it is convenient to write each clause using the *set-based notation*. So we will not repeat identical literals, will write in the same clause the comma instead of the disjunction $\vee$, will cancel all the occurrences of $\neg\neg$ and will use braces to enclose the literals of the clause. The three clauses of Example 12.6 will then be further simplified by writing $\{\neg Ux, Mx\}$, $\{\neg Px, \neg Sx, Hx\}$, and $\{\neg Bx, Syy, Sxy\}$.

The deviation of the clause-based language from the spoken language is the price we need to pay for developing the logical calculus.

Using the language of clauses, we will give in Theorem 14.1 a simple proof of a first version of Gödel completeness theorem. We will see in Theorem 16.10 that the fact that our logical calculus acts on clauses is not an essential limitation. In Theorem 16.11 we will extend the completeness theorem to predicate logic with equality.

12.3 Substitution of terms for variables

Definition 12.7. Assume a string E on the alphabet Σ. Writing

$$E(x_1, \ldots, x_n)$$

we wish to say that *variables of E belong to the set* $\{x_1, \ldots, x_n\}$. If we now have an n-tuple of terms $\mathbf{t} = (t_1, \ldots, t_n)$, we can *substitute in E each variable x_i by the corresponding term t_i*. The new string thus obtained is denoted by $E(\mathbf{t})$.

Example 12.8. Let $E(x, y, z)$ be the clause

$$\{Pf(x)g(b, y),\ Qg(y, a),\ Pyg(z, z)\}$$

and $\mathbf{t}$ be the triple of terms $(f(u), g(y, z), f(f(b)))$. Then $E(\mathbf{t})$ is the clause

$$\{Pf(f(u))g(b, g(y, z)),\ Qg(g(y, z), a), Pg(y, z)g(f(f(b)), f(f(b)))\}.$$

So if L_1, L_2, L_3 are the literals of E, we have $E(\mathbf{t}) = \{L_1(\mathbf{t}), L_2(\mathbf{t}), L_3(\mathbf{t})\}$.

Recall that the associativity of composition of functions states that $f(g(h)) = (f(g))(h)$. The following lemma shows that a similar associativity property holds for substitutions:

Lemma 12.9. *Given a string* $E = E(x_1, \ldots, x_n)$, *let* $\mathbf{t} = (t_1, \ldots, t_n)$ *be an* n-*tuple of terms, with* $t_i = t_i(y_1, \ldots, y_m)$, *for all* $i = 1, \ldots, n$. *Then for every* m-*tuple of terms* $\mathbf{r} = (r_1, \ldots, r_m)$ *the following identity holds:*

$$(E(\mathbf{t}))(\mathbf{r}) = E(\mathbf{t}(\mathbf{r})), \tag{12.1}$$

where $\mathbf{t}(\mathbf{r})$ *is an abbreviation of the* n-*tuple of terms* $(t_1(\mathbf{r}), \ldots, t_n(\mathbf{r}))$.

Proof. We proceed by induction on the number $l = 1, 2, \ldots$ of symbols in E, where, as mentioned above, each variable $x|\ldots|$ is counted as a *single* symbol. If $l = 1$ and E is not a variable symbol, then E remains unchanged for every substitution. If the unique symbol of E is a variable x_i, then $(x_i(\mathbf{t}))(\mathbf{r}) = t_i(\mathbf{r}) = x_i(\mathbf{t}(\mathbf{r}))$.

For the induction step, as $l > 1$, we can split E in two strings E' and E'' of length $< l$, so that each variable of E lies entirely either in E' or in E''. (In short, we want to avoid the possibility of a split inside of $x|\ldots|$.) The induction hypothesis holds for E' and E''. We write $E = E' \smile E''$ to indicate that E is the concatenation of E' and E''. Therefore we can write

$$(E(\mathbf{t}))(\mathbf{r}) = ((E' \smile E'')(\mathbf{t}))(\mathbf{r}) = (E'(\mathbf{t}) \smile E''(\mathbf{t}))(\mathbf{r}) =$$

$$= (E'(\mathbf{t}))(\mathbf{r}) \smile (E''(\mathbf{t}))(\mathbf{r}) = E'(\mathbf{t}(\mathbf{r})) \smile E''(\mathbf{t}(\mathbf{r})) = (E' \smile E'')(\mathbf{t}(\mathbf{r})),$$

which coincides with $E(\mathbf{t}(\mathbf{r}))$.   □

12.4 Herbrand universe

Definition 12.10. By a *CNF formula* of predicate logic we mean a finite conjunction of clauses, written as $\{C_1, \ldots, C_u\}$ in the set-based notation. Given a (finite or countably infinite) set S of clauses, by the *Herbrand universe* of S, denoted by H_S, we mean the set of all ground terms obtained from the constants and functions of S.

| If S does not contain any constant, we add one to it, for example c |

If there are no functions in S and S is finite, then H_S is a finite set. As soon as there is a function symbol in S, H_S is automatically infinite. For example if $S = \{\{Qax, Rf(b)ba, \neg Qbf(a)\}, \{Qxy, Rabx\}\}$ then $H_S = \{a, b, f(a), f(b), f(f(a)), f(f(b)), \ldots\}$.

Definition 12.11. Let $H = H_S$ be the Herbrand universe of a set S of clauses. Let K be a subset of H and $C = C(x_1, \ldots, x_n)$ a clause of S. Then the *instance* C/K *of* C *over* K is the following set of ground clauses:

$$C/K = \{C(\mathbf{g}) \mid \mathbf{g} = (g_1, \ldots, g_n) \in K^n\}.$$

One then defines the *instance* S/K *of* S *over* K as

$$S/K = \bigcup_{C \in S} C/K. \tag{12.2}$$

Example 12.12. Let S be a set of clauses, with the Herbrand universe

$$H = \{a, b, f(a), f(b), f(f(a)), f(f(b)), \ldots\}.$$

Let $K = \{a, b, f(a), f(b)\}$. Further, let $C = \{Ax, Bf(x), \neg Ab\}$ be a clause of S.

Instantiating C over K means listing the clauses that one obtains substituting the variable x with the terms from K, in all possible ways. All these clauses will be ground. Therefore:

- the substitution of x with a is the clause $\{Aa, Bf(a), \neg Ab\}$;
- the substitution of x with b is the clause $\{Ab, Bf(b), \neg Ab\}$;
- the substitution of x with $f(a)$ is the clause $\{Af(a), Bf(f(a)), \neg Ab\}$;
- the substitution of x with $f(b)$ is the clause $\{Af(b), Bf(f(b)), \neg Ab\}$.

12.5 Refutation

Each ground clause C is a disjunction of the literals $L_1, \ldots, L_t$ that do not contain variables. Depending on the possible world in which it is interpreted, each literal L_i becomes true or false, just as a literal of propositional logic acquires a truth value by an assignment. The identification

| atomic ground formula = propositional variable |

is a simple and clever trick in predicate logic. So for example, the set of ground clauses

$$\{\{Qc, Paf(b), Tbg(a, c)\}, \{\neg Qc, Paf(b)\}, \{Qc, \neg Paf(b)\}, \{\neg Tbg(a, c)\}\}$$

differs from the set of clauses

$$\{\{X, XI, XII\}, \{\neg X, XI\}, \{X, \neg XI\}, \{\neg XII\}\}$$

only in the way the variables are written. Identifying each finite set S of ground clauses of predicate logic with a set of clauses of propositional logic, DPP acts on S exactly as if S belonged to propositional logic, computing resolvents and resolvents of resolvents.

Definition 12.13. We say that a set S of clauses is *refutable* if the empty clause is derivable from a finite subset S' of S/H_S applying to S' the Davis-Putnam procedure, or simply, refuting S' using Definition 4.3.

Example 12.14. Let $S = \{\{\neg Ux, Mx\}, \{Ua\}, \{\neg Ma\}\}$ and

$$S' = S/H_S = \{\{\neg Ua, Ma\}, \{Ua\}, \{\neg Ma\}\}.$$

Then resolving the first two clauses of S' we obtain $\{Ma\}$, and resolving this clause with the third clause of S' we obtain the empty clause. This short refutation guarantees that, whatever possible world we conceive to interpret the symbols of S, in that world S cannot hold.

For example, in the world in which Ux means "x is ungulate", My means "y is left-handed", and a denotes "Alf", it cannot be the case that "every ungulate is left-handed, Alf is ungulate, Alf is not left-handed". And analogously, in another world in which Ux means "x is a man", My means "y is mortal", and a represents Andrew, it cannot be the case that "every man is mortal, Andrew is a man, Andrew is not mortal".

As in the propositional case, the above refutation can be represented economically by a graph; the only difference is that now we also have to state which instances are used to transform clauses into ground clauses:

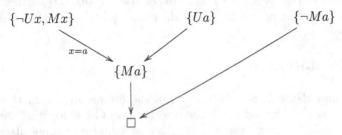

For centuries logic essentially finished here: today logic starts with an in-depth examination of the relation between the irrefutability of a set S of clauses and the existence of "possible worlds', called "models", in which S holds.

Exercises

1. Let $E = E(x, y, z)$ and $\mathbf{t} = \mathbf{t}(u, y, z)$ be as in Example 12.8. Let $\mathbf{r}$ be the triple of terms $(g(z, a), c, f(g(x, v)))$. Verify that

$$(E(\mathbf{t}))(\mathbf{r}) = E(\mathbf{t}(\mathbf{r})).$$

2. Verify that if S is a finite set of clauses, then S/H_S is finite iff S does not contain function symbols.

3. Let C be the clause $\{Px, Qay, Rzbx\}$. How many clauses are there in C/H_C?

4. For each set S of clauses and subsets $K_1 \subseteq K_2 \subseteq \ldots$ of H_S verify that

$$\bigcup \frac{S}{K_i} = \frac{S}{\bigcup K_i}.$$

5. Given the clause $\{Px, \neg Pf(a, y)\}$, list four elements of its Herbrand universe and instantiate the clause by these elements in all possible ways.

6. Given the set of clauses $S = \{\{\neg Ux, Mx\}, \{Ma\}, \{\neg Ua\}\}$, instantiate it over its Herbrand universe and verify that S/H is not refutable. *(If each man is mortal and Alf is mortal, why on earth should Alf be a man? Might it not be a dog?)*

7. Find a refutation of the set of clauses

$$S = \{\{\neg Fx, Px\}, \{\neg Cy, \neg Py\}, \{Fa\}, \{Ca\}\}$$

instantiating S over its Herbrand universe H, and then refuting (graphically) the set of ground clauses S/H as if they were clauses of propositional logic.

13

The Meaning of Clauses

13.1 Tarski semantics: types and models

The discovery of a non-Euclidean geometry ended a two thousand-year-old search for a refutation of the set of statements obtained by adding the negation of the Fifth Postulate (about parallel lines) to the remaining Euclidean axioms. This is an important example of a general fact: if there exists a possible world in which a set of statements is true, then there does not exist a refutation of this set of statements. The reverse implication is a fundamental result of logic, the completeness theorem of Gödel, for the proof of which we cannot content ourselves with a generic intuition. Instead, we have to learn to work with some fundamental concepts that will be defined in this chapter.

To start with, the intuitive notion of a "possible world" is defined as follows.

Definition 13.1. A *type* τ is a set of (constant, relation or function) symbols. A *model* M *of type* τ is a pair $(M, {}^*)$ where M is a nonempty set, called *the universe* of M, and $*$ is a function that contains τ in its domain, with the following properties:

- for each constant symbol $c \in \tau$, c^* is an element of M;

- for each function symbol $f \in \tau$ with n arguments, f^* is an n-ary function from M^n to M, in symbols $f^* \colon M^n \to M$;

- for each relation symbol $R \in \tau$ with m arguments, R^* is an m-ary relation in M, in symbols, $R^* \subseteq M^m$.

Let S be a CNF formula. Then the *type* of S is a set of (constant, relation or function) symbols of S. A model M is *suitable* for S if its type τ contains the type of S. One defines analogously the type of a term t and the property that M is suitable for t.

Mundici D.: Logic: a Brief Course.
DOI 10.1007/978-88-470-2361-1_13, © Springer-Verlag Italia 2012

Notation and tacit assumptions. From now on, all models will be tacitly assumed to be suitable for all clauses, terms and literals to which one will refer to. As we have already done in Definition 12.7 and in Lemma 12.9, for each term t we will write $t(x_1, \ldots, x_n)$ to state that the variables of t belong to the set $\{x_1, \ldots, x_n\}$. The same meaning has the notation $L(x_1, \ldots, x_n)$ and $C(x_1, \ldots, x_n)$, where L is a literal and C is a clause. We will write $\mathbf{x}$ as the abbreviation of the n-tuple $(x_1, \ldots, x_n)$. We will not use the implication connective and the existential quantifier until Chapter 16.

Definition 13.2. Assume given a term $t = t(x_1, \ldots, x_n)$, a model $\mathcal{M} = (M, ^*)$ and an n-tuple $\mathbf{m} = (m_1, \ldots, m_n) \in M^n$. By induction on the number of function symbols occurring in t, we define the element $t^{\mathcal{M}}[\mathbf{m}]$ of M as follows:

$$a^{\mathcal{M}}[\mathbf{m}] = a^*, \text{ for each constant symbol } a \tag{13.1}$$

(clearly $\mathbf{m}$ plays no rôle);

$$x_i^{\mathcal{M}}[\mathbf{m}] = m_i \; ; \tag{13.2}$$

and for each k-ary function symbol f and k-tuple $(t_1, \ldots, t_k)$ of terms,

$$(f(t_1, \ldots, t_k))^{\mathcal{M}}[\mathbf{m}] = f^*(t_1^{\mathcal{M}}[\mathbf{m}], \ldots, t_k^{\mathcal{M}}[\mathbf{m}]). \tag{13.3}$$

Given a p-tuple of terms $\mathbf{r} = (r_1(x_1, \ldots, x_n), \ldots, r_p(x_1, \ldots, x_n))$, one defines $\mathbf{r}^{\mathcal{M}}[\mathbf{m}] = (r_1^{\mathcal{M}}[\mathbf{m}], \ldots, r_p^{\mathcal{M}}[\mathbf{m}])$ in an analogous way.

So to give a meaning to a term $t = t(x_1, \ldots, x_n)$ in a model $\mathcal{M}$ we have to associate with $\mathcal{M}$ an n-tuple of elements $m_1, \ldots, m_n$ of its universe. When we defined $t(x_1, \ldots, x_n)$, we did not require that each variable x_i occurs in t. In reality it suffices to associate with $\mathcal{M}$ as many elements as there are variables that actually occur in t. In particular, when t is ground, we can associate with $\mathcal{M}$ the empty 0-tuple $\emptyset$, *that of course we will not write*. In this case, (13.3) takes the simplified form

$$(f(t_1, \ldots, t_k))^{\mathcal{M}} = f^*(t_1^{\mathcal{M}}, \ldots, t_k^{\mathcal{M}}). \tag{13.4}$$

For each p-tuple $\mathbf{s} = (s_1, \ldots, s_p)$ of ground terms we will use the notation

$$\mathbf{s}^{\mathcal{M}} = (s_1^{\mathcal{M}}, \ldots, s_p^{\mathcal{M}}). \tag{13.5}$$

Example 13.3. Let $\tau = \{c, s, f, g\}$ and let $\mathcal{M}$ be the model of type τ whose universe is the set $\mathbb{N} = \{0, 1, 2, \ldots\}$ of natural numbers, and in which c^*, s^*, f^*, g^* are respectively zero, the successor function, the addition and multiplication functions. Let $t(x, y)$ be the term given by $f(g(c, s(x)), f(x, y))$. Then $t^{\mathcal{M}}[(3, 7)] = (0 \cdot (3 + 1)) + (3 + 7) = 10$.

Exercise 13.4. Remaining with the same term $t(x, y)$ of the preceding example, construct a model $\mathcal{Q} = (\mathbb{Q}, ^\natural)$ of type τ having as universe the rational numbers in which $c^\natural, s^\natural, f^\natural, g^\natural$ are defined in such a way that $t^{\mathcal{Q}}[(1, 0)] = -1$.

The following lemma takes care of the relationship between substitution (a purely syntactical operation that transforms expressions into expressions) and the operation $g \mapsto g^{\mathcal{M}}$ (that transforms the ground term g into an element of the universe of the model $\mathcal{M}$):

Lemma 13.5. *Let* $\mathbf{t} = (t_1(x_1, \ldots, x_n), \ldots, t_p(x_1, \ldots, x_n))$, *and*

$$\mathbf{g} = (g_1, \ldots, g_n),$$

where each g_i *is a ground term. Then* $(\mathbf{t}(\mathbf{g}))^{\mathcal{M}} = \mathbf{t}^{\mathcal{M}}[\mathbf{g}^{\mathcal{M}}]$.

Proof. We proceed by induction on the number l of symbols in $\mathbf{t}$. If $l = 1$, $\mathbf{t}$ is a constant c or a variable x_i. If $t = c$, then $(c(\mathbf{g}))^{\mathcal{M}} = c^{\mathcal{M}} = c^* = c^{\mathcal{M}}[\mathbf{g}^{\mathcal{M}}]$. If $t = x_i$, then we have $(x_i(\mathbf{g}))^{\mathcal{M}} = g_i^{\mathcal{M}} = x_i^{\mathcal{M}}[\mathbf{g}^{\mathcal{M}}]$. For the induction step, to keep the notation simple, suppose that $p = 1$. Then $\mathbf{t}$ is of the form $f(\mathbf{h}) = f(h_1, \ldots, h_q)$, where each h_i is a term in the variables $x_1, \ldots, x_n$, for which the induction hypothesis holds. Therefore using Lemma 12.9 together with (13.3)-(13.4) we can write

$$(t(\mathbf{g}))^{\mathcal{M}} = ((f(\mathbf{h}))(\mathbf{g}))^{\mathcal{M}} = (f(\mathbf{h}(\mathbf{g})))^{\mathcal{M}} = f^*((\mathbf{h}(\mathbf{g}))^{\mathcal{M}}) =$$

$$= f^*(\mathbf{h}^{\mathcal{M}}[\mathbf{g}^{\mathcal{M}}]) = (f(\mathbf{h}))^{\mathcal{M}}[\mathbf{g}^{\mathcal{M}}] = \mathbf{t}^{\mathcal{M}}[\mathbf{g}^{\mathcal{M}}].$$

$\square$

13.2 Tarski semantics: clauses

Given a set S of clauses of type τ and a model $\mathcal{M} = (M, ^*)$ we are finally able to give a meaning to the statement "$\mathcal{M}$ satisfies S", in symbols $\mathcal{M} \models S$.

Definition 13.6. Suppose that $\mathcal{M}$ is suitable for a set S of clauses and that $P \in \tau$ is a k-ary relation symbol. Put $\mathbf{t} = (t_1(\mathbf{x}), \ldots, t_k(\mathbf{x}))$, where $\mathbf{x} = (x_1, \ldots, x_n)$, and suppose that $\mathbf{m} = (m_1, \ldots, m_n)$ is an n-tuple of elements of M. Then we define $\mathcal{M} \models S$ inductively as follows:

$$\mathcal{M}, \mathbf{m} \models P\mathbf{t} \quad \text{means} \quad \mathbf{t}^{\mathcal{M}}[\mathbf{m}] \in P^*; \tag{13.6}$$

$$\mathcal{M}, \mathbf{m} \models \neg P\mathbf{t} \quad \text{means} \quad \mathbf{t}^{\mathcal{M}}[\mathbf{m}] \notin P^*.$$

For each clause $C \in S$, where $C = C(x_1, \ldots, x_n)$, we write

$$\mathcal{M}, \mathbf{m} \models C \text{ iff for some literal } L \in C, \ \mathcal{M}, \mathbf{m} \models L$$

and

$$\mathcal{M} \models C \text{ iff for each } \mathbf{m} = (m_1, \ldots, m_n) \in M^n, \ \mathcal{M}, \mathbf{m} \models C.$$

Finally, $\mathcal{M} \models S$ means that $\mathcal{M} \models C$ for each $C \in S$. In other words,

$$\text{for each } C = C(\mathbf{x}) \in S \text{ and } \mathbf{m} \in M^n \ \mathcal{M}, \mathbf{m} \models L \text{ for some } L \in C. \tag{13.7}$$

This definition expresses rigorously the following imprecise intuition:

> $\mathcal{M} \models S$ *is supposed to say that each clause C of S becomes true if it is read with reference to the constant, relation and function symbols of $\mathcal{M}$, letting each variable of C range over the universe of $\mathcal{M}$.*

Whether $\mathcal{M}, m_1, \ldots, m_n$ satisfies C or not, only depends on the elements m_i of M associated with the variables of C. Thus in particular, when $\mathbf{t} = (t_1, \ldots, t_n)$ and each term t_i is ground we will write $\mathcal{M} \models P\mathbf{t}$ instead of $\mathcal{M}, \emptyset \models P\mathbf{t}$. Recalling now (13.5), we can write

$$\mathcal{M} \models P\mathbf{t} \quad \text{means} \quad \mathbf{t}^{\mathcal{M}} \in P^* \qquad \text{(when } \mathbf{t} \text{ is ground).} \qquad (13.8)$$

Definition 13.7. One says that a set S of clauses is *satisfiable* if there exists a model that satisfies it. Otherwise, S is said to be *unsatisfiable*. A clause C is a *(logical) consequence* of S if each model $\mathcal{M}$ that satisfies S also satisfies C (it is understood that $\mathcal{M}$ is suitable for both S and C).

Exercise 13.8. Verify that the set

$$S = \{\{Eaf(b)\}, \{\neg Eba\}, \{Ef(x)a, \neg Eaf(x)\}\}$$

of clauses is satisfied by the model $\mathcal{N} = (\mathbb{N},^*)$ where $a^* = 1, b^* = 0, f^*(n) = n + 1$, and E^* is the equality relation on $\mathbb{N}$. Find another model, preferably with fewer elements.

Exercise 13.9. Verify the satisfiability of each of the following sets of clauses:

(i) $\{\{Pa\}, \{Pg(g(a))\}, \{Pg(x), \neg Px\}\}$;

(ii) $\{\{Px, Qx\}, \{\neg Px, \neg Qx\}, \{Qa\}, \{\neg Qb\}\}$;

(iii) $\{\{Nx\}, \{\neg Nx, Nf(x)\}, \{\neg Nf(x), Nx\}\}$.

13.3 Instantiation, resolution and its correctness

The proof of Lemma 3.6 that sanctions the correctness of the resolution rule for propositional logic immediately guarantees the correctness of the resolution rule for predicate logic:

Corollary 13.10 (Correctness of ground resolution). *Let R be a finite set of ground clauses and $DPP(R)$ be the set of clauses generated from R by means of DPP. Then each clause $C \in DPP(R)$ is a logical consequence of R. In other words, if a model $\mathcal{M}$ satisfies R, then $\mathcal{M}$ also satisfies C.*

The next result confirms the intuition that if a ground clause G is obtained by an instantiation of a clause C, then G is a logical consequence of C. This intuition corresponds to the fact that the variables of C are tacitly bound by the universal quantifier $\forall$. The proof is a testing ground for checking our understanding of the notions introduced in this chapter:

Proposition 13.11 (Correctness of ground instantiation). *Let $C = C(x_1, \ldots, x_n)$ be a clause and let $\mathbf{g} = (g_1, \ldots, g_n)$ be an n-tuple of ground terms. If $\mathcal{M} \models C$, then $\mathcal{M} \models C(\mathbf{g})$ (always assuming that the model $\mathcal{M} = (M, *)$ is suitable for both clauses).*

Proof. Writing the clause C as $C = \{L_1, \ldots, L_u\}$ we have $C(\mathbf{g}) = \{L_1(\mathbf{g}), \ldots, L_u(\mathbf{g})\}$. Recalling (13.7), the assumption $\mathcal{M} \models C$ means that for all $\mathbf{m} \in M^n$ there exists a literal in C that is satisfied in $\mathcal{M}, \mathbf{m}$. In particular, $\mathcal{M}, \mathbf{g}^{\mathcal{M}} \models L$ for some $L \in C$. Suppose that $L = P\mathbf{t}$, where P is an k-ary relation and $\mathbf{t}$ is an k-tuple of terms. Therefore $\mathcal{M}, \mathbf{g}^{\mathcal{M}} \models P\mathbf{t}$.

By (13.6), $\mathbf{t}^{\mathcal{M}}[\mathbf{g}^{\mathcal{M}}] \in P^*$. By Lemma 13.5, $(\mathbf{t}(\mathbf{g}))^{\mathcal{M}} \in P^*$. As each term of the k-tuple $\mathbf{t}(\mathbf{g})$ is ground, on the account of (13.8) we can write $\mathcal{M} \models P(\mathbf{t}(\mathbf{g}))$, that is (Lemma 12.9), $\mathcal{M} \models (P\mathbf{t})(\mathbf{g})$. In other words, $\mathcal{M} \models L(\mathbf{g})$ and hence $\mathcal{M} \models C(\mathbf{g})$ as we wanted.

One deals with the case $L = \neg P\mathbf{t}$ the same way. $\qquad\square$

Combining this proposition with (12.2) we obtain immediately:

Corollary 13.12. *Let S be a set of clauses, with its Herbrand universe H. If $\mathcal{M} \models S$ then $\mathcal{M} \models S/H$.*

Exercises

In each of the following exercises verify that the conclusion is *not* a logical consequence of the premises, by transcribing the premises (that are above the line) and the negation of the conclusion into clauses, and subsequently finding a model that satisfies all these clauses.

1. Each pigeon is two-footed
   Bet is not a pigeon

   _____

   Bet is not two-footed

   Solution. The clauses $\{\neg Px, Tx\}$, $\{\neg Pb\}$, $\{Tb\}$ formalise the premises and the negation of the conclusion. Let $\mathcal{M} = (M, *)$ where $M = \{b^*\}$, $b^* = $ Napoleon, $C^* = \emptyset$, $B^* = M$. Then the model $\mathcal{M}$ satisfies all these clauses. Therefore it is not true that each model (a possible world) that satisfies the premises also satisfies the conclusion.

   It is important to note that when constructing a possible world for the premises and the negation of the conclusion we have complete freedom in interpreting the symbols. In particular, any resemblance of "Bet" to persons living or dead, bearing or not bearing this name, is purely coincidental. Actually, our present model $\mathcal{M}$ interprets as Napoleon the constant b (which abbreviates the string of symbols "Bet"). In older texts, to impress upon the students this abstraction and to avoid any confusion about Bet, this exercise could be given in the following dry form:

Each P is T
b is not P

---

b is not T

2. Each child loves his mother
   Beatrice loves her mother

Beatrice is a child

Solution. Premises: $\{\neg Cx, Lx\ m(x)\}, \{Lb\ m(b)\}$. Negation of the conclusion: $\{\neg Cb\}$. The model $\mathcal{M} = (M, *)$ in which we put

$$M = \{\text{Athena}\}, \quad b^* = \text{Athena}, \quad C^* = \emptyset, \quad L^* = \{(b^*,\ b^*)\}, \quad m^*(b^*) = b^*$$

satisfies the premises and the negation of the conclusion.
Note. To formalise "mother of ..." we have used the function symbol m. So in the model $\mathcal{M}$, m^* is a function, whence we are forced to specify who Athena's mother is, even if Athena is not a child. In our present model $\mathcal{M}$, Athena is her own mother. This is possible because no premise stipulates $m(x) \neq x$. The omission of this "axiom" is, possibly, a problem of the one who proposed these clauses. It is not our problem here. The problem of saying "the whole truth", for example about the equality, or about the natural numbers will be considered later.

3. Every combatant is mortal
   Alf is a combatant

Alf's mother is mortal

Solution. The premises and the negation of the conclusion are formalised by the set $S = \{\{\neg Cx, Mx\}, \{Ca\}, \{\neg Mm(a)\}\}$ of clauses. In this exercise everything that is supposed to be known about combatants, mortals, mothers, and Alf is contained in the set S of clauses. In particular, no clause of S is devoted to stating other facts that would seem obvious to us, like "Alf's mother differs from Alf", or "the mother of each mortal is mortal". We should not be distracted by all these extra hypotheses. Perhaps, using them we might succeed to deduce the thesis – but we have to work only on the set S of symbols given to us in this exercise. In a similar way, when we solve systems of equations we do not ask ourselves whether these equations faithfully represent intersecting geometrical figures or falling bodies.

We have to find a model of S. Each reader will find his preferred one. The model $\mathcal{Q} = (Q,^{\ddagger})$ here proposed is suggested by the Greek mythology. We put $Q = \{\text{Achilles, his mother, mother of his mother}, \ldots\}$, $a^{\ddagger} = \text{Achilles}$, $C^{\ddagger} = M^{\ddagger} = \{\text{Achilles}\}$, and $m^{\ddagger}(x) = \text{mother of } x$, for each $x \in Q$. The model satisfies S because Achilles is the unique combatant in the universe Q, is mortal, and his mother (the sea nymph Thetis) is not mortal. Therefore the conclusion is not a logical consequence of the premises.

4. Let $\mathcal{N} = (\mathbb{N},^{\dagger})$, where $\mathbb{N} = \{0, 1, 2, \ldots\}$, $C^{\dagger} = M^{\dagger} = \{0\}$, $a^{\dagger} = 0$ and $m^{\dagger} = $ the successor function. Verify that $\mathcal{N}$ satisfies the set S of Exercise 3.

5. A dog does not bite a dog
   Alf bites Blick
   Blick bites Alf

   _____

   Alf is not a dog

6. $\{\neg Ix, Pbx\}$
   $\{Pab\}$

   _____

   $\{\neg Ia\}$

7. Each fast talker promises everybody heaven and earth
   Dick promises everybody heaven and earth

   _____

   There is a fast talker

8. Each A is B
   each B is C
   each C is D

   _____

   some C is not B or some D is not A

9. $\{\neg Ax, Bx\}$
   $\{\neg By, Cy\}$
   $\{\neg Cz, Dz\}$

   _____

   $\{\neg Dt, At\}$

10. Prove the following statement:

    Let $\mathcal{M} = (M, *)$ be a model of type τ and let m be an element of M. Suppose that there exists a ground term t of type τ such that $t^{\mathcal{M}} = m$. Let $C(x)$ be a clause of type τ with a unique variable x. Then $\mathcal{M}, m \models C$ iff $\mathcal{M} \models C(t)$.

    Hint. Use the associativity of substitution together with Lemma 13.5.

11. State and prove the generalization of Lemma 13.5 in case $\mathbf{g}$ is an n-tuple of (not necessarily ground) terms.

14

Gödel's Completeness Theorem for the Logic of Clauses

14.1 Introduction

This fundamental theorem shows the equivalence of two at first sight different properties of a set of clauses S:

- satisfiability, or the existence of a model of S;
- coherence (= irrefutability = consistency), i.e., the impossibility of obtaining the empty clause when applying DPP to a finite subset of S/H_S.

Frege and Hilbert had contrasting ideas on these two notions. According to Frege the existence of a model of S is the definitive proof that S cannot be refuted: it does not have much sense to go on searching for other proofs of the coherence of S.

For example, from the axioms that speak about the commutativity and associativity of addition and multiplication of natural numbers, and of the distributivity of multiplication over addition, no one will ever obtain the empty clause, because these axioms obviously have a model.

Hilbert's view was the opposite: the coherence of a set S of axioms should be proved in a more down to earth, mechanical fashion, mimicking what DPP does when S is a set of clauses in propositional logic. Surely DPP does not take into account our trust that S has a model (or even, a preferred model) – a trust that turned out to be illusory more than once. DPP is just a simple mechanism for generating new clauses. Once the impossibility of obtaining the empty clause is ascertained, one can expect that some form of "model-building" yields a model of S.

It was Hilbert who first explicitly pointed out that refutations could be mathematical objects, and posed the *completeness problem of logic*. Using the language of these pages we can formulate the problem as follows:

Does each unsatisfiable set of clauses have a refutation?

Gödel's Completeness Theorem gives a positive answer to this problem.

Mundici D.: Logic: a Brief Course.
DOI 10.1007/978-88-470-2361-1_14, © Springer-Verlag Italia 2012

When S is a finite set of clauses and H' is a finite subset of its Herbrand universe, we will denote by

$$DPP(S/H')$$

the set of all clauses obtained from S/H' by applying DPP. As we have already noted on page 67, since S/H' is a finite set of ground clauses, there is no problem for DPP to handle such clauses as clauses of propositional logic.

14.2 Completeness and compactness

Theorem 14.1 (Gödel's Completeness Theorem). *Let S be a finite set of clauses of type τ, with its Herbrand universe H. Then the following statements are equivalent:*

(i) S is unsatisfiable;
(ii) S is refutable, in the sense that for some finite subset H' of H it holds $\square \in DPP(S/H')$.

Proof (according to Herbrand and Skolem).
$(ii) \Rightarrow (i)$ Assume by contradiction that S is both refutable and satisfiable, say $\mathcal{N} \models S$. From Corollary 13.12 it follows that $\mathcal{N} \models S/H$. A fortiori we have $\mathcal{N} \models S/H'$ and then by Corollary 13.10, $\mathcal{N} \models DPP(S/H')$. The two clauses that generate in $DPP(S/H')$ the empty clause have the form $\{Pt\}$ and $\{\neg Pt\}$ and are both satisfied in $\mathcal{N}$. By (13.8) we have both $\mathbf{t}^{\mathcal{N}} \in P^*$ and $\mathbf{t}^{\mathcal{N}} \notin P^*$, which is impossible.

$(i) \Rightarrow (ii)$ We will show that if S is not refutable then it is satisfiable. Let $H_1 \subseteq H_2 \subseteq \ldots$ be an increasing sequence of finite subsets of H with $H = \bigcup_n H_n$. For each fixed i, S/H_i is a finite set of clauses (because both S and H_i are finite). S/H_i is an input for DPP, whose variables are identified with the atomic ground formulas of S/H_i. By assumption, $\square \notin DPP(S/H_i)$. By Theorem 4.1 there exists an assignment α_i that satisfies (each clause of) S/H_i in the sense of propositional logic; in symbols, $\alpha_i \models_{\text{prop}} S/H_i$. As $S/H = S/\bigcup H_n = \bigcup S/H_n$, by the Compactness Theorem 6.2 there exists an assignment α that satisfies S/H in propositional logic; in symbols,

$$\alpha \models_{\text{prop}} S/H. \tag{14.1}$$

The domain of α is the set of atomic formulas of S/H. Let Pt be an atomic formula that occurs (possibly preceded by negation) in a clause of S/H. Then α assigns to Pt a truth value $\alpha(Pt) \in \{0, 1\}$. The assignment α suggests the following:

Construction of the model $\mathcal{M} = (M,^*)$ of S of type τ. To start with, we put $M = H$. For each n-ary predicate symbol $P \in \tau$, we define the relation $P^* \subseteq M^n$ putting for each $\mathbf{t} \in M^n = H^n$,

$$\mathbf{t} \in P^* \text{ iff } \alpha(Pt) = 1. \tag{14.2}$$

In other words, $\mathbf{t} \in P^*$ iff $\alpha \models_{prop} P\mathbf{t}$. In particular, if the ground atomic formula $P\mathbf{t}$ is not in the domain of α we have $\mathbf{t} \notin P^*$. For each constant $a \in \tau$ we put $a^* = a$. For each k-ary function symbol $f \in \tau$ we define the function $f^* \colon M^k \to M$ putting $f^*(t_1, \ldots, t_k) = f(t_1, \ldots, t_k)$. This completes the construction of $\mathcal{M}$.

Recalling the notation (13.4), and arguing by induction on the number of functions symbols in the term $h \in H$ we see that

$$h^{\mathcal{M}} = h. \tag{14.3}$$

Final claim. $\mathcal{M} \models S$.

For each clause $C = C(x_1, \ldots, x_n) \in S$ and n-tuple $\mathbf{g} \in M^n = H^n$ we want to prove that $\mathcal{M}, \mathbf{g} \models C$. We write $C = \{L_1, \ldots, L_u\}$, so $C(\mathbf{g}) = \{L_1(\mathbf{g}), \ldots, L_u(\mathbf{g})\}$. As $C(\mathbf{g})$ is an element of S/H, it follows from (14.1) that

$$\alpha \models_{prop} C(\mathbf{g}), \quad \text{that is,} \quad \alpha \models_{prop} L(\mathbf{g}) \text{ for some literal } L \in C.$$

Suppose that L has the form $\neg P\mathbf{t}$ (the case $L = P\mathbf{t}$ is similar). We have $\alpha \models_{prop} (\neg P\mathbf{t})(\mathbf{g})$, and therefore $\alpha \models_{prop} \neg P(\mathbf{t}(\mathbf{g}))$. From (14.2) we obtain $\mathbf{t}(\mathbf{g}) \notin P^*$. Applying (14.3) to the sequence $\mathbf{t}(\mathbf{g})$ of ground terms we can write $(\mathbf{t}(\mathbf{g}))^{\mathcal{M}} \notin P^*$. By Lemma 13.5 we obtain $\mathbf{t}^{\mathcal{M}}[\mathbf{g}^{\mathcal{M}}] \notin P^*$. By (13.6) we can write $\mathcal{M}, \mathbf{g}^{\mathcal{M}} \models \neg P\mathbf{t}$. Again by (14.3) we obtain $\mathcal{M}, \mathbf{g} \models \neg P\mathbf{t}$, that is, $\mathcal{M}, \mathbf{g} \models L$ and hence $\mathcal{M}, \mathbf{g} \models C$, as we wanted. $\quad\square$

Exercise 14.2. The two statements (i) and (ii) of the Completeness Theorem 14.1 are also equivalent to each of the following two statements:

(iii) for *each* sequence $H_1 \subseteq H_2 \subseteq \ldots$ of finite subsets of H with $H = \bigcup H_n$, there exists some i such that $\square \in DPP(S/H_i)$;

(iv) for *some* sequence $H_1 \subseteq H_2 \subseteq \ldots$ of finite subsets of H with $H = \bigcup H_n$, there exists some i such that $\square \in DPP(S/H_i)$.

Theorem 14.3 (Gödel Compactness Theorem). *Let S be a countably infinite set of clauses of type τ, with its Herbrand universe H. Then the following statements are equivalent:*

(i) S is unsatisfiable;

(ii) some finite subset S' of S is unsatisfiable;

(iii) for some finite subset S' of S and finite subset H' of H we have $\square \in DPP(S'/H')$.

Proof. By the Completeness Theorem $(iii) \Leftrightarrow (ii)$. Trivially, $(ii) \Rightarrow (i)$. The proof that $(i) \Rightarrow (iii)$ proceeds by supposing that for no finite subset S' of S and finite subset H' of H we have $\square \in DPP(S'/H')$. Then for each pair

(S', H') the model-building procedure of propositional logic provides an assignment $\alpha_{S',H'}$ satisfying S'/H' in propositional logic. It is easy to see that

$$\frac{S}{H} = \bigcup \left\{ \frac{S'}{H'} \mid S' \subseteq S, \quad H' \subseteq H, \quad S', H' \text{ finite} \right\}. \tag{14.4}$$

Therefore S/H is finitely satisfiable in propositional logic. The Compactness Theorem for propositional logic provides an assignment $\alpha \models_{\text{prop}} S/H$. From α we construct a model of S by means of the same technique as the one used in the proof of the Completeness Theorem 14.1.

Therefore S is satisfiable. We have proved $\neg(iii) \Rightarrow \neg(i)$, that is, $(i) \Rightarrow (iii)$. $\square$

14.3 Comments on the Completeness Theorem

The steps of DPP that, having as input the finite subset S/H' of S/H, arrive at the empty clause, produce nothing else but strings of symbols of the alphabet Σ. The resulting sequence of symbols is, however, an ultimate, mathematically convincing "proof" that the statement formed by S does not hold in any "possible world" $\mathcal{M}$, and hence S is unsatisfiable.

But what to think if, no matter how we choose finite sets $H_1 \subseteq H_2 \subseteq \ldots$ with $H = \bigcup H_n$, and apply DPP to each S/H_i we *never* get the empty clause? Gödel's Completeness Theorem states that this sequence of failed attempts of finding the empty clause is constructing a model of S.[1]

On the other hand, anybody who announces having proved the unsatisfiability of S, perhaps also the Martian quoted on the previous pages, equipped with extraterrestrial deductive powers, is warned that also our meticulous reasoner DPP will arrive sooner or later at the same result, by producing the empty clause.

Gödel's Completeness Theorem does not preclude the discovery of new techniques of proving theorems that are faster or more efficient than the ones we use today.

And in fact, in the subsequent sections we will describe some tools (for example, the equality predicate, the nonclausal formulas) used in the mathematical practice. But in essence, "proofs" boil down to the time-honoured manipulations of our logical calculus, namely instantiation and resolution: no one is allowed to draw conclusions that cannot be drawn from this calculus.

As we have seen, Gödel's Completeness Theorem states that the unsatisfiability of S (a condition that calls for a galaxy of "possible worlds" to be explored by S) is equivalent to the existence of a refutation of S – meaning that for some H_i the empty clause belongs to $DPP(S/H_i)$. So the contrast

---

[1] Recall the failed attempt of Saccheri to prove the incompatibility of the postulates of Euclid with the negation of the Fifth Postulate: page after page, he was in fact describing a model of non-Euclidean geometry.

between Frege and Hilbert regarding the satisfiability of S doesn't look so drastic, after all: the former asked for the existence of a model of S, while the latter asked for the irrefutability of S; and the Completeness Theorem states that these two conditions are equivalent.

Actually, there would be no contrast at all if, just as the unsatisfiability of a set S of clauses is always certified by a refutation, also the satisfiability of S could be taken care of by some sort of "satisfiability certificate" – say, a "mechanical procedure" $DPP^\natural$ that, on input S, terminates in a finite number of steps iff S is satisfiable. Then, by simultaneously launching DPP and $DPP^\natural$, after a finite number of steps precisely one of them would terminate, allowing us to decide mechanically whether S is satisfiable or unsatisfiable. After all, this is the status of satisfiability and unsatisfiability in propositional logic.

A series of fundamental results of 20th century mathematical logic, the coverage of which asks for a second course in logic, (i) gave a convincing definition of a "mechanical procedure that terminates in a finite number of steps" and (ii) proved the nonexistence of a mechanical procedure that terminates in a finite number of steps iff S is satisfiable. This is the Turing-Church Theorem on the undecidability of predicate logic, answering Hilbert's fundamental *Decision Problem (Entscheidungsproblem)*.

Exercises

Refutational Method

In each of these exercises the conclusion is a consequence of the premises. *Deduce* the conclusion from the premises in a finite number of purely formal steps using the following *refutational method* that realises the proof of Gödel theorem:

(i)   Formalise the premises and the negation of the conclusion, thus obtaining a finite set of clauses S with its Herbrand universe H.

(ii) Instantiate S over a finite subset H' of H; for brevity try to write a refutation with a few resolvents compared to those produced by DPP: only the computer has enough time and patience for carry out completely DPP(S/H').

(iii) Check whether DPP(S/H') produces the empty clause. Otherwise enlarge H' to a subset $H'' \supseteq H'$ and return to step (ii). Only look for the few instantiations that are really needed to obtain the empty clause.

1. Every combatant is mortal
   Ach is a combatant
   the mother of every combatant is mortal
   _____

   the mother of Ach is mortal

Hint. Formalising the premises and the negation of the conclusion we obtain the set $R = \{\{\neg Cx, Mx\}, \{Ca\}, \{\neg Mz, Mm(z)\}, \{\neg Mm(a)\}\}$ of clauses. All information about combatants, mortals, mothers and Ach that we can use to solve this exercise is contained in the set of clauses R.

We realise that there does not exist a "possible world" $\mathcal{M}$, possibly a mythological one, that is a model of R. This intuition is confirmed by the following refutation of R: instantiating R over the subset $\{a\}$ of the Herbrand universe of R we obtain an unsatisfiable set $R/\{a\}$ of clauses of propositional logic. If $\mathcal{M}$ existed, it would have to satisfy $R/\{a\}$, together with every clause obtained by means of DPP. But among these clauses there is the empty clause.

The following graph represents a refutation of the premises and of the negation of the conclusion; all instantiations are over a:

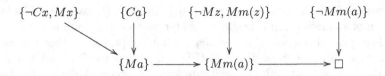

2. Dog does not bite a dog
   Alf bites Blick

   ―――――――――――――――――――

   at least one among Alf and Blick is not a dog

Solution. The following graph represents a refutation of the premises and of the negation of the conclusion:

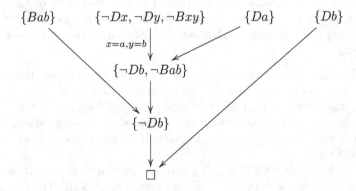

3. The tardy bird does not catch the worm
   Alf is a bird
   Alf catches Bic
   Bic is a worm

   ―――――――――――――――――――

   Alf is not tardy

4. Each primate is a vertebrate
   the father of each primate is a primate
   Arg is a primate

   ---

   the father of Arg is a vertebrate

5. $\{\neg Px, \neg Ry, Txy\}$
   $\{\neg Rz, Tzb\}$
   $\{Ra\}$
   $\{Pa\}$

   $\{Taa, Tab\}$

   Hint. Negating the conclusion means producing two clauses $\{\neg Taa\}$ and $\{\neg Tab\}$.
   Note. Not all six clauses of this exercise are necessary for a refutation.

6. Every pianist admires every violinist
   Every violinist admires Bill
   Alice is both a pianist and a violinist

   Alice admires herself and Bill

7. $\{\neg Fx, Pm(x)\}$
   $\{\neg Cy, \neg Pm(y)\}$

   $\{\neg Fa, \neg Ca\}$

8. In Canto XXVI of Dante's *Inferno* the devil gives a lesson of logic to Saint
   Francis who is unfairly bringing count Guido da Montefeltro to paradise.
   In verses 118-119 one finds two premises:[2]

   > *One may not be absolved without repentance,*
   > *nor repent and wish to sin concurrently.*

   In the preceding verses we find the third premise, under the form of a
   self-denouncement:

   > *Guido da Montefeltro wants to commit a sin.*

   Finally, there is a doctrinal premise that is so obvious that Dante does
   not need to announce it explicitly:

   > *Everybody who is not absolved goes to hell.*

   Using the constant $c =$ count Guido da Montefeltro and the predicates
   $Rx =$ "x repents", $Ay =$ "y is absolved", $Wz =$ "z wants to commit
   a sin", and $Hu =$ "u goes to hell", deduce from the four premises the
   diabolic conclusion that Guido da Montefeltro goes to hell.

   ---

   [2] Dante Alighieri, *The Inferno*, translation by Robert Hollander.

Solution (instantiating each variable by *c*):

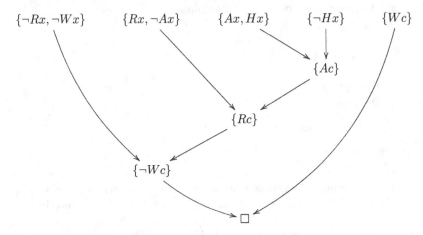

9. Find in the preceding exercise another refutation, starting from the resolvent $\{\neg Rc\}$ of $\{\neg Rc, \neg Wc\}$ and $\{Wc\}$.

Satisfiability/Unsatisfiability

For each of the following exercises write in clauses the premises and the negation of the conclusion; then informally assess whether the conclusion is a consequence of the premises. If this is not the case, construct a model (preferably with a few elements) for the premise P and the negation N of the conclusion. If instead the conclusion is a consequence of the premise, by appropriately instantiating P and N over the Herbrand universe, find the empty clause using the refutational method. Gödel theorem assures us that this approach succeeds one way or another.

1. Every hare fears every wolf: $\{\neg Hx, \neg Wy, Fxy\}$
   Bic fears nobody: $\{\neg Fbz\}$
   Ark is a wolf: $\{Wa\}$

   ───────────────────────────────

   Bic is not a hare: $\{\neg Hb\}$

2. Every hare fears any fox that chases it
   Bic is a hare

   ───────────────────────────────

   Bic fears some fox

3. $\{\neg Px, \neg Ey, Mm(x)m(y)\}$
   $\{\neg Mm(a)m(b)\}$

   _____

   $\{\neg Pb, \neg Ea\}$

4. $\{\neg Px, \neg Ey, Mm(x)m(y)\}$
   $\{\neg Mm(a)m(b)\}$

   _____

   $\{\neg Pa, \neg Eb\}$

5. The mass of every flea is smaller than the mass of every elephant
   The mass of Alf is not smaller than the mass of Bet
   Bet is an elephant

   _____

   Alf is not a flea

6. $\{\neg Ix, \neg Axf(y)\}$
   $\{If(z)\}$

   _____

   $\{\neg Af(a)a\}$

7. Those who are uneducated cannot defend themselves against touts
   Biago cannot defend himself against himself

   _____

   Biagio is uneducated or is not a tout

8. In chapter VI of the short story *The Death of Ivan Ilych* by Lev Tolstoy, the protagonist is confronted with the syllogism

   Caius is a man, men are mortal, therefore Caius is mortal

   he had studied in a logic manual that was in vogue at his times. Comment on the considerations of Ivan Ilych and assess the plausibility of his observations. Here is the relevant fragment:[3]

   > Ivan Ilych saw that he was dying, and he was in continual despair. In the depth of his heart he knew he was dying, but not only was he not accustomed to the thought, he simply did not and could not grasp it. The syllogism he had learnt from Kiesewetter's Logic: "Caius is a man, men are mortal, therefore Caius is mortal," had always seemed to him correct as applied to Caius, but certainly not as applied to himself. That Caius – man in the abstract – was mortal, was perfectly correct, but he was not Caius, not an abstract man, but a creature quite, quite separate from all others. He had been little Vanya, with a mamma and a papa, with Mitya and Volodya, with the toys, a coachman and a nurse, afterwards with Katenka and with all the joys, griefs, and delights of

   _____

   [3] L. Tolstoy, *The Death of Ivan Ilych*, translation by Louise and Aylmer Maude.

childhood, boyhood, and youth. What did Caius know of the smell of
that striped leather ball Vanya had been so fond of? Had Caius kissed
his mother's hand like that, and did the silk of her dress rustle so for
Caius? Had he rioted like that at school when the pastry was bad? Had
Caius been in love like that? Could Caius preside at a session as he
did? "Caius really was mortal, and it was right for him to die; but for
me, little Vanya, Ivan Ilych, with all my thoughts and emotions, it's
altogether a different matter. It cannot be that I ought to die. That
would be too terrible."
Such was his feeling:
"If I had to die like Caius I would have known it was so. An inner voice
would have told me so, but there was nothing of the sort in me and I and
all my friends felt that our case was quite different from that of Caius.
and now here it is!" he said to himself. "It can't be. It's impossible! But
here it is. How is this? How is one to understand it?"

a) As a mathematical exercise, this syllogism is perfectly valid for a generic man, like Caius, but cannot be applied to Ivan Ilych, who is not a
generic man and has his own personal life story which Caius cannot
claim as his own. Caius can pass away at any moment, but by a similar
approach as the one discussed in the solution of Exercise 3 on page
76, by adding appropriate axioms specific for Ivan Ilych, his mother
and his relatives, the conclusion of the syllogism could be turned into
the conclusion that Ivan Ilych will continue to live for several years.

b) Just as the conjunction "if", also the adjective "every" has a more
restricted meaning in mathematics than in the natural language. One
thing is to say "every crow is black", another is to say "every natural number has a successor", and still another thing is to say "every
nonempty set of natural numbers has a least element". Such simple
symbolic manipulations as instantiation and resolution just fall short
of taking control of all the facets of the adjective "every".

15

Equality Axioms

15.1 Introduction

The possibility of enriching the logical calculus with the equality symbol is not a trifling matter. This symbol is found everywhere in mathematics, and makes sense in every Tarskian model. This is not the case for the other mathematical relations. For example, the relation "is bigger than" makes no sense for lines, the relation "belongs to" makes no sense for natural numbers, etc.

Therefore we enrich our alphabet with the new binary relation symbol $\approx$. Later on, we will write $=$ instead of $\approx$, but for now the risk of confusion is too big. For convenience, we will write $x \approx y$ instead of $\approx xy$. We will define a *model with equality* $\mathcal{M} = (M,^*)$ where the equality symbol $\approx$ is interpreted as the equality relation on M. Therefore

$$\approx^* \quad \text{coincides with the relation} \quad \{(x,y) \in M^2 \mid x = y\}.$$

It may happen that the equality symbol occurs in a satisfiable set S of clauses, but no model with equality satisfies S:

Example 15.1. The set S constructed from three clauses

$$\{a \approx b\}, \{Pa\}, \{\neg Pb\}$$

is satisfiable, for example, by a model $\mathcal{M} = (M,^*)$ whose universe has two elements a^*, b^*, the binary relation symbol $\approx$ is interpreted as the set M^2, (therefore $x \approx^* y$ for every $x, y \in M$), and the predicate P is interpreted as the singleton set $\{a^*\}$.

On the other hand, S is not satisfiable by any model $\mathcal{N} = (N,^\natural)$ with equality. In fact, in $\mathcal{N}$ we have that $a^\natural$ and $b^\natural$ are the same element, and therefore it is impossible that $a^\natural \in P^\natural$ and $b^\natural \notin P^\natural$.

Mundici D.: Logic: a Brief Course.
DOI 10.1007/978-88-470-2361-1_15, © Springer-Verlag Italia 2012

To extend the Completeness Theorem to all clauses containing the equality symbol, we need to extend the logical calculus. Fortunately this extension does not call for new "deduction rules" beyond instantiation and resolution. Rather, we will just add the few clauses – the "equality axioms" – that constitute the outcome of thousand years of reflections on the properties of equality.

15.2 Axiomatisation of the equality

To begin, recall three *equality axioms*, that we write in the usual mathematical language:

- (Reflexivity) $\forall x \ (x \approx x)$;
- (Symmetry) $\forall x \forall y \ (x \approx y \to y \approx x)$;
- (Transitivity) $\forall x \forall y \forall z \ ((x \approx y \land y \approx z) \to x \approx z)$.

For each function symbol $f(x_1, \ldots, x_n)$ and relation $P x_1 \cdots x_m$, using the abbreviations $f(\mathbf{x})$ and $P\mathbf{x}$, we also introduce the following *congruence axioms*:

$$\forall x_1 \cdots \forall x_n \forall x_1' \cdots \forall x_n'((x_1 \approx x_1' \land \ldots \land x_n \approx x_n') \to f(\mathbf{x}) \approx f(\mathbf{x'}))$$

and

$$\forall x_1 \cdots \forall x_m \forall x_1' \cdots \forall x_m' \ ((x_1 \approx x_1' \land \ldots \land x_m \approx x_m') \to (P\mathbf{x} \to P\mathbf{x'})).$$

Consequently, when we have a set S of clauses in which there appears the equality symbol, the logical calculus adds to S three clauses

$$\{x \approx x\}, \ \ \{\neg x \approx y, \ y \approx x\}, \ \ \{\neg x \approx y, \ \neg y \approx z, \ x \approx z\},$$

and for each function symbol f and relation symbol P that appears in S, it also adds the clauses

$$\{\neg x_1 \approx x_1', \ldots, \neg x_n \approx x_n', \ f(x_1, \ldots, x_n) \approx f(x_1', \ldots, x_n')\}$$

$$\{\neg x_1 \approx x_1', \ldots, \neg x_m \approx x_m', \ \neg P x_1, \ldots, x_m, \ P x_1', \ldots, x_m'\}.$$

Instead of $\neg x \approx y$ we will write $x \not\approx y$.

Theorem 15.2. *Let S be a set of clauses. Let $S^\approx$ be the set of clauses obtained by adding to S the clauses of the equality axioms and the congruence axioms for the function and relation symbols of S. Then the following conditions are equivalent:*

(i) S is satisfiable by a model with equality;

(ii) $S^\approx$ is satisfiable.

Proof. (i) $\Rightarrow$ 2 Let $\mathcal{N} = (N,^\sharp)$ be a model with equality that satisfies S. Clearly, the relation $\approx^\sharp$ is an equivalence relation on the universe N. Further, the congruence axioms hold in $\mathcal{N}$ for each function and relation symbol that occurs in S. Therefore $\mathcal{N} \models S^\approx$.

(ii) $\Rightarrow$ 1 Let $\mathcal{M} = (M,^*)$ be a model of $S^\approx$. From the fact that $\mathcal{M}$ satisfies the first three axioms, we obtain that the relation $\approx^* \subseteq M^2$ is an equivalence relation on M. Let N be the set of its equivalence classes. For each $x \in M$ let $\langle x \rangle \in N$ be its equivalence class. We will construct a model $\mathcal{N} = (N,^\sharp)$ with equality that satisfies S. Therefore we define $\approx^\sharp$ as the equality relation on N. For each n-ary function symbol f putting

$$f^\sharp(\langle a_1 \rangle, \ldots, \langle a_n \rangle) = \langle f^*(a_1, \ldots, a_n) \rangle, \quad (a_i \in M) \tag{15.1}$$

we have a correct[1] definition of a function $f^\sharp \colon N^n \to N$. This is so because $\mathcal{M}$ satisfies the congruence axiom for f. Analogously, putting

$$(\langle b_1 \rangle, \ldots, \langle b_m \rangle) \in P^\sharp \text{ iff } (b_1, \ldots, b_m) \in P^* \ (b_i \in M) \tag{15.2}$$

we define an m-ary relation $P^\sharp$ in N. This concludes the definition of $\mathcal{N}$. By construction $\mathcal{N}$ is a model with equality. From (15.1) for each term $t(x_1, \ldots, x_k)$ and k-tuple $\mathbf{a} = (a_1, \ldots, a_k) \in M^k$, by induction on the number of function symbols in t, we obtain

$$\langle t^\mathcal{M}[\mathbf{a}] \rangle = t^\mathcal{N}[\langle \mathbf{a} \rangle], \qquad \text{where } \langle \mathbf{a} \rangle \text{ stands for } (\langle a_1 \rangle, \ldots, \langle a_k \rangle). \tag{15.3}$$

Let L be a literal of the form $P\mathbf{t}$, where $\mathbf{t} = t(x_1, \ldots, x_k)$ (the case $L = \neg P\mathbf{t}$ is analogous), and let $\mathbf{m}$ be an element of M^k. Using (15.1)–(15.3) we see that

$$\mathcal{M}, \mathbf{m} \models L \quad \text{iff} \quad \mathcal{N}, \langle \mathbf{m} \rangle \models L.$$

Therefore, for each clause $C = C(x_1, \ldots, x_k)$ of S and $\mathbf{m} \in M^k$ we have

$$\mathcal{M}, \mathbf{m} \models C \text{ iff } \mathcal{N}, \langle \mathbf{m} \rangle \models C.$$

As $\mathcal{M} \models C$ for each clause $C \in S$, we conclude that $\mathcal{N} \models S$. □

Example 15.3 (Continuation of Example 15.1). The set $S = \{\{a \approx b\}, \{Pa\}, \{\neg Pb\}\}$ becomes unsatisfiable as soon as we add the congruence axiom

$$\{x \not\approx y, \neg Px, Py\}$$

for P. Let us instantiate this axiom replacing x by a and y by b. We then obtain the set $\{\{a \approx b\}, \{Pa\}, \{\neg Pb\}, \{a \not\approx b, \neg Pa, Pb\}\}$ of clauses that is easily seen to be refutable.

| From now on we abolish the heavy symbol $\approx$ and write $=$ instead |

---

[1] i.e., independent of the choice of the representantive x in the class $\langle x \rangle$.

Exercises

1. Formalise the following statements as clauses with equality:

   a) if Clementine is the maternal grandmother of Luisa, then she is also the mother of Filippo and of Marta;

   b) there are two second prize winners and one first prize winner;

   c) either the earth is the unique planet inhabited by mathematicians, or at least two mathematicians live on different planets.

2. Verify by means of a refutation the unsatisfiability of the set S of Example 15.3.

3. Describe a model for the set

   $$S = \{\{a = b\}, \{f(a) = b\}, \{f(b) \neq a\}\}$$

   of clauses. Verify that S does not have any model with equality.

4. Using the resolution method prove that the conclusion is a logical consequence of the premise:

   $$\forall x \forall y (x = a \lor y = a)$$
   _____
   $$f(a, b) = f(b, a)$$

   Solution:

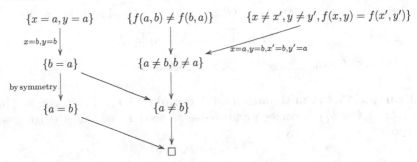

   Note. "by symmetry" means that one should take a resolvent of the clause $\{b = a\}$ with an instantiation of the clause that expresses the symmetry of equality.

5. For each example check whether the conclusion is a logical consequence
   of the premise. If not, find a model with equality for the premise and the
   negated conclusion; if yes, prove it using the refutational method:

   a) $a \neq b$
      $\forall x (x = a \lor x = b)$
      $\forall x \; x = f(x)$
      $f(c) = b$
      _____

      $c = b$

   b) $a \neq b$
      $\forall x (x = a \lor x = b)$
      $\forall x \; x \neq f(x)$
      _____

      $\exists x \; f(f(x)) = x$

   c) $\forall x (x = a \lor x = b)$
      _____

      $\exists x \; f(f(x)) = x$

16

The Predicate Logic $\mathcal{L}$

16.1 Introduction

In this chapter, a bit more dense than the other ones, we describe a logic, denoted by $\mathcal{L}$, and known as "predicate logic" or "elementary logic", or also "first-order logic." The syntax of $\mathcal{L}$ is perfectly tailored for the language of mathematics, which usually avoids clauses.

Let Σ be the alphabet defined at the beginning of Chapter 12. Adding the equality symbol we have the alphabet $\Sigma^=$. We update Definition 12.5 of an atomic formula decreeing that also every string $t_1 = t_2$ (where t_1, t_2 are terms) is an atomic formula.

Definition 16.1. The formulas (of $\mathcal{L}$) are the strings over $\Sigma^=$ given by the following inductive definition:

- each atomic formula is a formula;
- if F and G are formulas, then $\neg F, (F \vee G), (F \wedge G), (F \to G)$, are formulas;
- if F is a formula and x is a variable, then $(\exists x F)$ and $(\forall x F)$ are formulas.

For each formula F the *unique readability* principle holds, that allows us to uniquely decompose F into its constituents. The proof is a variant of that of Theorem 7.3. This principle also allows us to define the *free occurrences* of a variable x in a formula F as follows:

- each occurrence of each variable in an atomic formula is free;
- the free occurrences of a variable in a negated formula $\neg F$ are exactly its free occurrences in F;
- the free occurrences of a variable in a formula of the type $(F \vee G), (F \wedge G), (F \to G)$, are the free occurrences in F together with the free occurrences in G;
- the variable x does not have a free occurrence in any formula $(\exists x F)$ or $(\forall x F)$, while each other variable y has the same free occurrences as in F.

Mundici D.: Logic: a Brief Course.
DOI 10.1007/978-88-470-2361-1_16, © Springer-Verlag Italia 2012

An occurrence of x in F is called *bound* if it is not free. A variable z is called *free* in F if it has a free occurrence; z is called *bound* in F if it has a bound occurrence.

The example of the formula $((\exists z Pz) \wedge Qz)$ shows that a variable can be both free and bound. Writing $F(x_1, \ldots, x_n)$ we mean to say that the free variables of the formula F are included in the set $\{x_1, \ldots, x_n\}$.

Definition 16.2. A *statement* is a formula in which all occurrences of the variables are bound.

From now on the word "model" stands for "model with equality". Further, all models will be suitable for all formulas to which they refer.

For simplicity we will omit outer parentheses and will apply the precedence rules already introduced for propositional logic.

Definition 16.3 (Extension of Definition 13.6 by induction on the number of connectives and quantifiers in F). Let $F(x_1, \ldots, x_n)$ be a formula $\mathcal{M} = (M, *)$ a model, and $\mathbf{m} = (m_1, \ldots, m_n)$ an n-tuple of elements of M.

– if F is an atomic formula, of the form Pt, then $\mathcal{M}, \mathbf{m} \models F$ (read: "$\mathcal{M}$ with $\mathbf{m}$ *satisfies* F") means that $\mathbf{t}^{\mathcal{M}}[\mathbf{m}]$ is an element of P^*;

– if F is $u = v$, $\mathcal{M}, \mathbf{m} \models F$ means that $u^{\mathcal{M}}[\mathbf{m}] = v^{\mathcal{M}}[\mathbf{m}]$;

– if F is a negated formula, $\neg G$, then $\mathcal{M}, \mathbf{m} \models \neg G$ means that it is not true that $\mathcal{M}, \mathbf{m} \models G$;

– if F is a conjunction or disjunction or an implication of two formulas H and K, then we stipulate that:

   – $\mathcal{M}, \mathbf{m} \models H \wedge K$ means that $\mathcal{M}, \mathbf{m} \models H$ and $\mathcal{M}, \mathbf{m} \models K$;

   – $\mathcal{M}, \mathbf{m} \models H \vee K$ means that $\mathcal{M}, \mathbf{m} \models H$ or $\mathcal{M}, \mathbf{m} \models K$;

   – $\mathcal{M}, \mathbf{m} \models H \rightarrow K$ means that if $\mathcal{M}, \mathbf{m} \models H$ then $\mathcal{M}, \mathbf{m} \models K$;

– if F is of the form $\forall x G(x_1, \ldots, x_n, x)$, then $\mathcal{M}, m_1, \ldots, m_n \models F$ means that for each $m \in M$ we have $\mathcal{M}, m_1, \ldots, m_n, m \models G$;

– if F is of the form $\exists x G(x_1, \ldots, x_n, x)$, then $\mathcal{M}, m_1, \ldots, m_n \models F$ means that there exists $m \in M$ such that $\mathcal{M}, m_1, \ldots, m_n, m \models G$.

This definition may appear pedantic because it just explains the intuition according to which $\mathcal{M}$ with $\mathbf{m}$ satisfies F if from the reading of F, after substituting the constant, function and predicate symbols by respective elements, functions and relations of $\mathcal{M}$, and after substituting each variable x_i by the element m_i, and interpreting the quantifiers "for all" and "there exists" on the universe of $\mathcal{M}$, one obtains a true statement. The point is that with this imprecise intuition we will not succeed to advance in the study of predicate logic: for our proofs we will need to work by induction on the number of symbols in F using the unique reading property of F.

We further note that this definition also gives meaning to formulas such as $\exists z\, a = b$, or $\forall x \exists x Pxx$ that upon first reading may appear ungrammatical.

Definition 16.4. We say that the formula $F(x_1, \ldots, x_n)$ is *satisfiable* if there exists a model $\mathcal{M}$ with an n-tuple $\mathbf{m}$ of elements of its universe M such that $\mathcal{M}, \mathbf{m} \models F$.

Formula $G(x_1, \ldots, x_n)$ is *(logically) equivalent* to F, in symbols $F \equiv G$, if for every model $\mathcal{M}$ suitable for both formulas and for every $\mathbf{m} \in M^n$ we have $\mathcal{M}, \mathbf{m} \models F$ iff $\mathcal{M}, \mathbf{m} \models G$.

A statement E is a *logical consequence* of the statements $E_1, \ldots, E_m$ if every model $\mathcal{M}$ of $E_1 \wedge \ldots \wedge E_m$ is a model of E. It is understood that $\mathcal{M}$ is suitable for both statements.

From these definitions we immediately have:

Proposition 16.5. *Given statements $E_1, \ldots, E_m, E$ the following conditions are equivalent:*

(i) E *is a logical consequence of* $E_1, \ldots, E_m$;

(ii) *the statement* $(E_1 \wedge \ldots \wedge E_m) \to E$ *is a tautology, i.e., it is satisfied in every model suitable for it;*

(iii) *the statement* $E_1 \wedge \ldots \wedge E_m \wedge \neg E$ *is unsatisfiable.*

So the notion of the logical consequence can be reduced to the notion of unsatisfiability.

16.2 Transformation of formulas in PNF

To apply to $\mathcal{L}$ the logical calculus prepared in the preceding chapters, we will transform each statement F into an appropriate set $\mathcal{S}_F$ of clauses such that F is satisfiable iff $\mathcal{S}_F$ is.

We first consider the following proposition, the proof of which is a good exercise for checking our understanding of the usefulness of Definition 16.3:

Proposition 16.6. *Let F, G, H, K be formulas. Then we have the following equivalences:*

(i) $F \to G \equiv \neg F \vee G$;

(ii) *if* $F \equiv G$, *then* $\neg F \equiv \neg G$;

(iii) *if* $F \equiv G$ *and* $H \equiv K$, *then* $F \wedge H \equiv G \wedge K$ *and* $F \vee H \equiv G \vee K$;

(iv) *if* $F \equiv G$, *then* $\exists x F \equiv \exists x G$ *and* $\forall x F \equiv \forall x G$;

(v) $\neg \exists x F \equiv \forall x \neg F$; $\neg \forall x F \equiv \exists x \neg F$;

(vi) *if x does not have any free occurrence in G, then* $(\forall x F) \wedge G \equiv \forall x (F \wedge G)$, $(\exists x F) \wedge G \equiv \exists x (F \wedge G)$, $(\forall x F) \vee G \equiv \forall x (F \vee G)$, $(\exists x F) \vee G \equiv \exists x (F \vee G)$.

The transformation of a formula F into a set of clauses is based on the following preliminary result:

Theorem 16.7. *Every formula F is equivalent to a PNF formula[1], that is, a formula $\mathcal{P}_F$ of the form*

$$Q_1 x_1 \cdots Q_k x_k\, G, \quad \text{in short,} \quad \mathbf{Qx}\, G, \tag{16.1}$$

of the same type and with the same free variables as F, that enjoys the following properties:

(i) G is quantifier-free and does not contain the implication symbol $\rightarrow$;

(ii)   for all $i = 1, \ldots, k$, $Q_i \in \{\exists, \forall\}$;

(iii) $x_i \neq x_j$ for $i \neq j$.

Proof. Applying Proposition 16.6(i) we can assume that the connective $\rightarrow$ does not occur already in F. We now proceed by induction on the number n of connectives and quantifiers in F, using the unique reading property of F. If $n = 0$, then F is an atomic formula and it suffices to put $\mathcal{P}_F = F$. For the induction step we have the following cases:

Case 1. $F = \neg K$. Using the induction hypothesis and Proposition 16.6(ii) we can write $F \equiv \neg \mathcal{P}_K$. Repeatedly applying Proposition 16.6(v) we obtain the desired formula $\mathcal{P}_F$.

Case 2. $F = H \vee K$. Using the induction hypothesis and Proposition 16.6(iii) we can write $F \equiv \mathcal{P}_H \vee \mathcal{P}_K$. Using the notation of (16.1) we write $\mathcal{P}_H = \mathbf{Q'y}A$ and $\mathcal{P}_K = \mathbf{Q''z}B$.

We note that every free variable of A does not have a bound occurrence in A, but some free variable of $\mathbf{Q'y}A$ might have a bound occurrence in $\mathbf{Q''z}B$, and vice versa. Therefore we rewrite all bound variables $\mathbf{y}$ of $\mathcal{P}_H$ and $\mathbf{z}$ of $\mathcal{P}_K$ using "new" variables $\mathbf{u}$ and $\mathbf{w}$ (that is, not occurring in $\mathcal{P}_H$ and $\mathcal{P}_K$).

We denote by $A_{\mathbf{u}}$ and $B_{\mathbf{w}}$ the result of this transformation of the formulas A and B. Clearly, $\mathbf{Q'y}A \equiv \mathbf{Q'u}A_{\mathbf{u}}$ and $\mathbf{Q''z}B \equiv \mathbf{Q''w}B_{\mathbf{w}}$. Applying repeatedly Proposition 16.6(vi) we see that $\mathbf{Q'uQ''w}A_{\mathbf{u}} \vee B_{\mathbf{w}}$ is the desired prenex normal form for F.

Case 3. $F = H \wedge K$. It is similar to Case 2.

Case 4. $F = \exists y L$. Applying the induction hypothesis and Proposition 16.6(iv) we can write $F \equiv \exists y \mathcal{P}_L$.

If y does not occur in $\mathcal{P}_L$, the quantification $\exists y$ does not play any rôle in $\mathcal{P}_L$ and we can put $\mathcal{P}_F = \exists y \mathcal{P}_L$, or simply $\mathcal{P}_F = \mathcal{P}_L$.

If y has a bound occurrence in $\mathcal{P}_L$, (and therefore, as we have noted, does not have any free occurrence) substituting y in $\mathcal{P}_L$ by a new variable we obtain a new PNF for L, and proceed as in the preceding case. Also in this case the quantification $\exists y$ does not play any rôle in $\mathcal{P}_L$.

---

[1] PNF stands for "prenex normal form".

There remains the case in which y is free (and therefore has no bound occurrence) in $\mathcal{P}_L$. Then the formula $\exists y\mathcal{P}_L$ is the desired PNF for F.

Case 5. $F = \forall yL$. Similar to Case 4.

Direct inspection shows that F and $\mathcal{P}_F$ have the same type. This completes the proof. □

Example 16.8. Omitting for the sake of readability some parentheses, one obtains the prenex normal form of the formula $(\exists x\forall yPxy)\vee((\exists xQx)\wedge(\forall yQy))$ by writing $(\exists x\forall yPxy) \vee ((\exists zQz) \wedge (\forall wQw))$, and then

$$\exists x\forall y\exists z\forall w(Pxy \vee (Qz \wedge Qw)). \tag{16.2}$$

The steps that transform F into $\mathcal{P}_F$ are completely mechanical and $\mathcal{P}_F$ is just a bit longer than F. This holds in general.

Whenever the commutativity of the binary connectives allows it, it is convenient to give precedence to the existential quantifier: so for example it is more convenient to transform the statement $\forall xPx\wedge\exists yQy$ into $\exists y\forall x(Px\wedge Qy)$ than into $\forall x\exists y(Px \wedge Qy)$. Attention! $\exists y\forall xPxy$ is not equivalent to $\forall x\exists yPxy$.

16.3 Skolemisation

Suppose now that F is a *statement*. Therefore its prenex normal form $\mathcal{P}_F = Q_1x_1 \cdots Q_kx_kG$ satisfies conditions (i)-(iii) of Theorem 16.7 and the variables of $\mathcal{P}_F$ are included in the set $\{x_1,\ldots,x_k\}$.

Suppose that some quantifier Q_i is $\exists$.

Case 1. $Q_1 = \exists$. Then a *Skolemisation step* consists of substituting x_1 in G by a new constant a, and transforming $\mathcal{P}_F$ into the statement

$$sk(\mathcal{P}_F) = Q_2x_2 \cdots Q_kx_kG(a, x_2,\ldots,x_k).$$

Case 2. The first existential quantifier of $\mathcal{P}_F$ is preceded by n universal quantifiers, $\mathcal{P}_F = \forall x_1 \cdots \forall x_n\exists x_{n+1}\mathbf{Q}\mathbf{x}G$. Then letting f be a new n-ary function symbol, a *Skolemisation step* of $\mathcal{P}_F$ produces the statement

$$sk(\mathcal{P}_F) = \forall x_1 \cdots \forall x_n\mathbf{Q}\mathbf{x}\, G(x_1,\ldots,x_n, f(x_1,\ldots,x_n), x_{n+2},\ldots,x_k), \tag{16.3}$$

obtained by substituting the variable x_{n+1} in $\mathcal{P}_F$ by the term $f(x_1,\ldots,x_n)$.

So a Skolemisation step eliminates in each statement that satisfies conditions (i)–(iii) of Theorem 16.7 the first existential quantifier from the left and produces a new statement that satisfies these three conditions.

Assuming there are m existential quantifiers in $\mathcal{P}_F$, applying m Skolemisation steps we have a finite sequence of statements

$$\mathcal{P}_F \mapsto sk(\mathcal{P}_F) \mapsto sk(sk(\mathcal{P}_F)) \mapsto sk(sk(sk(\mathcal{P}_F))) \mapsto \ldots$$

that transform the statement $\mathcal{P}_F$ into the statement $\forall z_1 \cdots \forall z_{k-m} H$, where H does not contain quantifiers, and the variables z_j are the universally quantified variables x_i of $\mathcal{P}_F$ listed in the same order. We call $\forall z_1 \cdots \forall z_{k-m} H$ the *Skolemisation* of $\mathcal{P}_F$.

Example 16.9. The statement $\exists x \forall y Pxy$ is Skolemised to $\forall y Pay$. By the same token $\forall x \exists y Gxy$ becomes $\forall x Gxf(x)$. The statement $\exists x \exists y \forall z \forall w \exists v \exists t Pxv \wedge Qvwz \wedge Pyyxvt$ becomes $\forall z \forall w Paf(z,w) \wedge Qf(z,w)wz \wedge Pbbaf(z,w)g(z,w)$. Applying two Skolemisation steps, statement (16.2) is transformed into

$$\forall y \forall w (Pay \vee (Qf(y) \wedge Qw)). \tag{16.4}$$

Theorem 16.10 (Skolem). *Given a statement F of type τ suppose that its prenex normal form $\mathcal{P}_F$ has m existential quantifiers. Let $\mathcal{S}_F$ be the formula obtained from $\mathcal{P}_F$ by means of m Skolemisation steps. Then we have:*

(i)   every model of $\mathcal{S}_F$ is also a model of F;

(ii)   if the model $\mathcal{M}$ of type τ satisfies F, then $\mathcal{S}_F$ has a model $\mathcal{M}'$ with the same universe as $\mathcal{M}$ and with the same interpretation of the symbols of F;

(iii) F is satisfiable iff $\mathcal{S}_F$ is satisfiable.

Proof. It suffices to consider a single Skolemisation step. Moreover, by Theorem 16.7 we can suppose that F is already in PNF, i.e., $F = \mathcal{P}_F$. We will consider only Case 2 of which Case 1 is a trivial variant. Let $sk(\mathcal{P}_F)$ be as in (16.3).

(i) Let $\mathcal{N} = (N, *)$ be a model of $sk(\mathcal{P}_F)$. By Definition 16.3 this means that for every n-tuple $d_1, \ldots, d_n$ of the elements of N it holds

$$\mathcal{N}, d_1, \ldots, d_n, f^*(d_1, \ldots, d_n) \models \mathbf{Qx}\ G(x_1, \ldots, x_k).$$

A fortiori, for every n-tuple $d_1, \ldots, d_n$ of the elements of N there exists an element $d \in N$ such that $\mathcal{N}, d_1, \ldots, d_n, d, \models \mathbf{Qx}\ G(x_1, \ldots, x_k)$. Again by Definition 16.3 this means that $\mathcal{N} \models \forall x_1 \cdots \forall x_n \exists x_{n+1} \mathbf{Qx} G$, that is, $\mathcal{N} \models \mathcal{P}_F$.

(ii) Let $\mathcal{M} = (M, \natural)$ be a model of $\mathcal{P}_F$. By Definition 16.3, for every n-tuple $e_1, \ldots, e_n$ of the elements of M there exists an element $e \in M$ such that

$$\mathcal{M}, e_1, \ldots, e_n, e, \models \mathbf{Qx}\ G(x_1, \ldots, x_k).$$

Using the Axiom of Choice[2] we can write e as $e = s(e_1, \ldots, e_n)$ for some function $s: M^n \to M$. Therefore for every $e_1, \ldots, e_n \in M$ we have:

$$\mathcal{M}, e_1, \ldots, e_n, s(e_1, \ldots, e_n) \models \mathbf{Qx}\ G(x_1, \ldots, x_k).$$

---

[2] This is the statement that the Cartesian product of a nonempty set of nonempty sets is nonempty.

Let $\natural\hspace{-0.3em}\natural$ be an extension of the function $\natural$ obtained by adding to the domain τ of $\natural$ a new n-ary function symbol f and putting $f^{\natural\hspace{-0.3em}\natural} = s$. Let $\mathcal{M}' = (M, \natural\hspace{-0.3em}\natural)$. Then for every $e_1, \ldots, e_n \in M$ we have

$$\mathcal{M}', e_1, \ldots, e_n, f^{\natural\hspace{-0.3em}\natural}(e_1, \ldots, e_n) \models \mathbf{Qx}\ G(x_1, \ldots, x_n, x_{n+1}, x_{n+2}, \ldots, x_k)$$

and therefore $\mathcal{M}' \models \forall x_1 \cdots \forall x_n \mathbf{Qx}\ G(x_1, \ldots, x_n, f(x_1, \ldots, x_n), x_{n+2}, \ldots, x_k)$, that is, $\mathcal{M}' \models sk(\mathcal{P}_F)$.

(iii) It follows immediately from (i) and (ii). $\qquad\qquad\square$

16.4 Completeness, compactness and nonstandard models

As we have seen, the Skolemisation of statement F produces the statement $\mathcal{S}_F = \forall z_1 \cdots \forall z_{k-m} H$, that enjoys the property of *equisatisfiability* established in Theorem 16.10(iii), with H quantifier-free. The equivalences of propositional logic (Theorem 9.4) continue to hold unchanged for all formulas of predicate logic. Therefore we can obtain from H an equivalent CNF formula H'. Applying Proposition 16.6(iv) we obtain $\forall z_1 \cdots \forall z_n H \equiv \forall z_1 \cdots \forall z_n H'$. Since the Skolemisation process eliminated all the existential quantifiers, eliminating $\forall z_1 \cdots \forall z_n$ and rewriting the clauses of H' in a set-based notation, we finally obtain a finite set $\mathcal{C}_F$ of clauses.

For example, the statement (16.4) becomes $\forall y \forall w((Pay \lor Qf(y)) \land (Pay \lor Qw))$ and then $\{\{Pay, Qf(y)\}, \{Pay, Qw\}\}$.

Noting that the transformations $F \mapsto \mathcal{P}_F \mapsto \mathcal{S}_F \mapsto \mathcal{C}_F$ of Theorems 16.7 and 16.10 can be derived mechanically, our complete logical calculus developed in Theorems 14.1 and 15.2 now extends to an equally complete logical calculus for predicate logic with equality $\mathcal{L}$:

Theorem 16.11 (Gödel's Completeness Theorem). *Let F be a statement of $\mathcal{L}$ and $\mathcal{C}_F$ the set of clauses obtained from F. Further, let $\mathcal{C}_F^{\equiv}$ be the set of clauses obtained by adding to $\mathcal{C}_F$ the equality and congruence axioms for all the relation and function symbols of $\mathcal{C}_F$. Let H be the Herbrand universe of $\mathcal{C}_F^{\equiv}$. Then F is unsatisfiable iff $\square \in DPP(\mathcal{C}_F^{\equiv}/H')$ for some finite subset H' of H.*

Recalling Theorem 14.3 we now immediately obtain the following two corollaries:

Corollary 16.12 (Gödel Compactness Theorem for $\mathcal{L}$). *Let Θ be a countably infinite set of statements of the logic $\mathcal{L}$. Then Θ is satisfiable iff each finite subset of Θ is satisfiable.*

Corollary 16.13 (Löwenheim, 1915 for finite Θ; Skolem, 1920, 1928 for the general case). *Let Θ be a finite or countably infinite set of statements of $\mathcal{L}$. If Θ has a model, then Θ has a model whose universe is finite or countably infinite.*

From this result it follows that *no finite or countably infinite set Θ of statements can characterise the ring structure $\mathcal{R}$ of the real numbers* – in the sense that $\mathcal{R}' \models \Theta$ iff $\mathcal{R}'$ is isomorphic to $\mathcal{R}$. In fact, if $\mathcal{R} \models \Theta$, then Θ also has a finite or countably infinite model $\widetilde{\mathcal{R}}$, and obviously $\widetilde{\mathcal{R}}$ is not isomorphic to $\mathcal{R}$.

Things are no better for the natural numbers:

Corollary 16.14. *Let $\mathcal{N} = (\mathbb{N}, ^\natural)$ be the structure of the natural numbers. Suppose that the type τ of $\mathcal{N}$ contains at least a symbol a for the zero, a symbol s for the successor function, and a binary predicate symbol G, where one reads Gxy as "x is larger than y". Let Θ be a finite or countably infinite list of statements of $\mathcal{L}$ such that $\mathcal{N} \models \Theta$. Then $\mathcal{N}' \models \Theta$ for some countable model $\mathcal{N}' = (N', ^*)$ that, in addition to the 'standard' elements a^*, $s^*(a^*)$, $s^*(s^*(a^*)), \ldots$, also contains a 'nonstandard' element larger than all standard elements.*

Proof. Let τ' be the type obtained by adding to τ a constant c. Let Θ' be the set obtained by adding to Θ the following new statements:

$$Gca, \ Gcs(a), \ Gcs(s(a)), \ldots \tag{16.5}$$

Each finite subset of Θ' is satisfiable – for example, in the model $(\mathbb{N}, ^*)$ of type τ' in which the function $*$ is an extension of $\natural$ to type τ', and c^* is a sufficiently large number. This follows from the assumption that $\mathcal{N} \models \Theta$. By Corollary 16.12, Θ' is satisfiable. By Corollary 16.13, Θ' has a countable model $\mathcal{N}' = (N', ^*)$. This model satisfies in particular all the statements (16.5). Therefore in $\mathcal{N}'$ the constant c is interpreted as a nonstandard element c^*. $\qquad\square$

Let τ be a type containing symbols for addition, multiplication, zero, one, and for an order relation $\leq$. Let $\mathcal{A}$ be a set of statements of type τ satisfied by the real numbers. Then $\mathcal{A}$ has a model $\mathcal{R}^*$ containing the *infinitesimals*, that is, elements $\epsilon > 0$ but smaller than every number of the form $\frac{1}{1+\cdots+1}$, and thus smaller than every "standard" real > 0. One proves this as in the previous corollary, using the compactness of $\mathcal{L}$: it is sufficient to add to τ a new constant c, and then add to $\mathcal{A}$ the following list of axioms:

$$0 < c, \ c < 1, \ c \cdot (1+1) < 1, \ c \cdot (1+1+1) < 1, \ldots$$

Clearly $\mathcal{R}^*$ does not satisfy the Archimedean principle, according to which for every two "quantities" $x, y > 0$ there exists some $n \in \mathbb{N}$ such that $nx > y$.

Exercises

Transformation into clauses

1. Transform into a set of clauses each of the following formulas, in which various parentheses are omitted for the sake of readability:

   a) $(\forall x Px) \wedge (\exists x Qx)$;

   b) $(\exists x \forall y Pxy) \rightarrow (\neg \forall w \forall y Pwy)$;

   c) $(\forall x Px) \rightarrow ((\forall x Qx) \rightarrow \forall x Sx)$;

   d) $((\forall x Px \rightarrow \forall y Dy) \rightarrow \forall y Sy) \rightarrow \forall y Ty$;

   e) $(\forall x \exists y Dxy \wedge \forall y \exists z Byz) \rightarrow \forall x \exists y \exists z (Cxy \rightarrow Dyz)$;

   f) $\forall x (Ax \wedge \forall y (Pxy \rightarrow \exists z (f(x,z) = y))) \rightarrow \forall y Pty$;

   g) $(\forall x (Cx \rightarrow \exists y Axy)) \rightarrow \neg \exists x (Cx \rightarrow \forall y Ayx)$.

2. Formalise in clauses using the equality predicate:

   a) there exist at least four elementary particles;

   b) for each pair of distinct points there passes exactly one line;

   c) each line passes by at least two distinct points;

   d) two lines have the same direction iff they are equal or do not have any point in common;

   e) for each external point of a line there passes exactly one parallel.

3. Let Wx stand for "x won the polls".
   Using when necessary the equality predicate, write down as a set of clauses each of the following phrases:

   a) at least one person won the polls;

   b) at most one person won the polls;

   c) exactly one person won the polls;

   d) at least two people won the polls;

   e) at most two people won the polls;

   f) only two people won the polls;

   g) if somebody won the polls, Luigi won the polls;

   h) if somebody won the polls, he was the only one.

4. Suppose that Fxy means "x is father of y", Mxy means "x is mother of y", c denotes Carlo and d denotes Damiano. The following statements express familiar relations between Carlo and Damiano. What are these relations?

   a) $\exists x \exists y (Mxc \wedge Mxd \wedge Fyc \wedge Fyd)$;

   b) $\exists x \exists y \exists z (Fxy \wedge Fxz \wedge Myc \wedge Fzd)$;

   c) $\exists x (Fcx \wedge Mxd)$;

   d) $\exists x (Fcx \wedge Fxd)$.

5. Using the predicates F and M and the constants c and d of the previous exercise, formalise in clauses the following phrases:

   a) Carlo is the paternal grandfather of Damiano;

   b) Carlo and Damiano are brothers;

   c) Carlo and Damiano do not have the same parents;

   d) Carlo and Damiano have the same maternal grandmother;

   e) the father of Damiano is a unique son.

6. Write in clauses the following statements, using the predicate Lxy per "x likes to play with y", and the predicates Vz and Pu for, respectively, "z is a violinist" and "u is a pianist":

   Every violinist either likes to play with every pianist or does not like to play with any pianist or likes to play with some violinist.

7. Transform into a set of clauses *the negation of the following phrase:*

   If for every innocent there is a judge who acquits him, then for every guilty there is a judge who condemns him.

   By convention "guilty = not innocent" and "condemned = not acquitted".

8. Write in clauses:

   a) fortune helps all courageous people and also helps some non courageous;

   b) if everything is moved by something, then there exists something that is moved only by itself;

   c) if there exists at least one extraterrestrial, then there exist at least two;

   d) if there was exactly one winner of the lottery, then either Carlo did not win it or Beatrice did not in it.

9. Write in clauses and find a model that satisfies these clauses:

   a) every Manichean condemns all the apostates or does not condemn any;

   b) every striker admires some goal keeper, but there are some goal keepers who are not admired by all strikers;

   c) if everything is moved by something, then there exists something that moves everything and is not moved by anything else.

Logical consequence

1. In the following exercises, we will write $P \not\models C$ to denote the fact that the conclusion C (written at the right) is not a consequence of the premise P.[3] Find in each case a model that satisfies the premise and the negation N of the conclusion.

   a) $\forall x Px \rightarrow \forall x Qx \not\models \forall x (Px \rightarrow Qx)$;

   b) $\exists x (Px \rightarrow Qx) \not\models (\exists x Px) \rightarrow \exists x Qx$;

   c) $\exists x Px \leftrightarrow \exists x Qx \not\models \exists x (Px \leftrightarrow Qx)$;

   d) $\forall x Px \leftrightarrow \forall x Qx \not\models \forall x (Px \leftrightarrow Qx)$;

   e) $\exists x Px \wedge \exists x Qx \not\models \exists x (Px \wedge Qx)$.

2. In these exercises the conclusion C is a consequence of the premise P. Formalise in clauses P and the negation of C. Produce a refutation of the clauses thus obtained, instantiating them appropriately over their Herbrand universe, and applying DPP.

   a) $\exists x \forall y Pxy \models \exists x Pxf(x)$;

   b) $\exists x (Px \wedge Qx) \models (\exists x Px) \wedge \exists x Qx$;

   c) $\exists x (Px \vee Qx) \models (\exists x Px) \vee \exists x Qx$;

   d) $\forall x \forall y Pxy \models \forall y \forall x Pxy$;

   e) $\exists x \exists y Pxy \models \exists y \exists x Pxy$;

   f) $\forall x \forall y Pxy \models \forall x Pxx$;

   g) $\forall x Pxf(x) \models \forall x \exists y Pxy$;

   h) $\forall x (Px \leftrightarrow Qx) \models (\exists x Px) \leftrightarrow (\exists x Qx)$;

   i) $\exists x Px \rightarrow \exists x Qx \models \exists x (Px \rightarrow Qx)$.

---

[3] This is an abuse of the notation because $\models$ has already been chosen to signify that a model satisfies a formula in Tarskian semantics. Nevertheless the context allows us to distinguish these two uses of this symbol.

Solution of the last exercise:

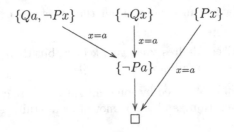

3. In each of the following exercises formalise in clauses the premise and the negation of the conclusion. Let S be the set of clauses thus obtained. If the conclusion is a consequence of the premise produce a refutation of S. Otherwise find a model.

a) $\forall x \exists y Pxy \models \exists x \exists y Pf(x, y)y$;

b) $\forall x \forall y Pxy \models \forall y Pyy$;

c) $\forall x(Px \vee Qx) \models \exists x(Px \vee Qx)$;

d) $\forall x \exists y Pxy \models \exists y Pyy$;

e) $\forall x \forall y(y = x) \models \forall y(y = a)$;

f) $\exists x Px \vee \forall x Qx \models \forall x(Px \vee Qx)$.

4. Verify the following tautologies, showing that each of them is a consequence of the equality axioms. Use the refutational method, negating the formula and writing this negated formula as a set of clauses and using appropriate equality axioms.

a) $\forall x \forall y[x = y \rightarrow (Txyx \rightarrow Tyxy)]$;

b) $\forall x \forall x' \forall y \forall y' [(x = x' \wedge y = y') \rightarrow g(f(x, y)) = g(f(x', y'))]$;

c) $\forall x \forall y \forall z[(x = y \wedge y = z \wedge Pxz) \rightarrow Pyy]$;

d) $\forall x \forall y[x = y \rightarrow (Pxy \rightarrow Pxx)]$;

e) $\forall x \forall x' \forall y \forall y' [(x = x' \wedge y = y') \rightarrow f(f(x, y), x) = f(f(x', y'), x)]$.

5. Determine which of the following statements are tautologies. If a statement E is a tautology, produce a refutation of $\neg E$, after having transformed $\neg E$ into a set of clauses. If E is not a tautology, find a model of $\neg E$, preferably with few elements.

   a) $\forall x \exists y (Pxy \rightarrow Pyx)$;

   b) $\forall x \forall y [(Px \leftrightarrow Py) \rightarrow x = y]$;

   c) $(\exists x \forall y (x = y)) \rightarrow (\forall x Px \vee \forall x \neg Px)$;

   d) $(\exists x Pxx) \rightarrow ((\forall y \forall w \neg Pyw) \rightarrow \exists t Ptt)$.

6. Prove using the refutational method that the conclusion is a logical consequence of the premise:

   $$\forall x \forall y (x = a \vee x = y)$$
   $$\overline{\forall u f(u) = f(f(u))}$$

   Abbreviated solution:

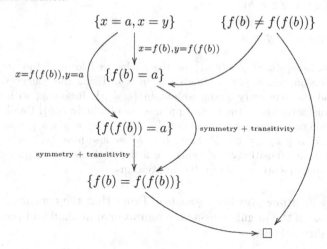

7. In each of the following exercises write in clauses the premises and the negation of the conclusion. Find out whether the conclusion is a consequence of the premises. If this is the case, using the equality axioms give a refutation of the set P of premises jointly with the negation N of the conclusion. Otherwise, construct a model with equality for P and N.

a) $\forall x \forall y\ x = y$

$$\overline{\quad \forall u \forall v\ f(u) = f(v) \quad}$$

b) $\forall x \forall y\ x \neq y$

$$\overline{\quad \forall u \forall v\ (f(u) = f(v) \wedge \neg g(u) = g(v)) \quad}$$

c) $\forall x \forall y\ (x = a \vee y = b)$
   $f(a, b) = f(b, a)$

$$\overline{\quad \forall z\ f(a, z) = f(z, a) \quad}$$

8. Prove without using the Completeness Theorem that the conjunction of the first three equality axioms cannot be refuted.[4]

9. Prove that the conclusion is a consequence of the premises using the refutational method and appropriate equality axioms:

$g(a, b) = g(b, a)$
$\forall z\ (z = a \vee z = b)$

$$\overline{\quad \forall x\ g(a, x) = g(x, a) \quad}$$

Hint. Let $\{g(a, c) \neq g(c, a)\}$ be the Skolemised form of the negation of the conclusion, where $p \neq q$ stands for $\neg p = q$. Using the congruence axiom for g and the symmetry axiom we obtain $\{c \neq a\}$. Resolving with an appropriate instantiation of the second premise we obtain $\{c = b\}$. Resolving with the congruence axiom for g we obtain $\{a \neq a, g(a, c) = g(a, b)\}$, and therefore by reflexivity $g(a, c) = g(a, b)$. From $\{c = b\}$ we also have $\{g(b, a) = g(c, a)\}$. Applying the transitivity and using the first premise we have $\{g(a, c) = g(c, a)\}$, and consequently we obtain the empty clause.

10. *Why does one plus one equal two.* Prove that the conclusion is a consequence of the premises using the refutational method and the appropriate equality axioms:

$\forall x \forall y\ f(x, s(y)) = s(f(x, y))$
$\forall x\ f(x, o) = x$

$$\overline{\quad f(s(o), s(o)) = s(s(o)) \quad}$$

---

[4] For Frege this exercise probably would be futile, as the satisfiability of the equality axioms is evident (indeed, their validity in each model with equality is evident).

Solution. Instantiating both premises we get

$$\{f(s(o), s(o)) = s(f(s(o), o))\} \text{ and } \{f(s(o), o) = s(o)\}.$$

From the last premise, resolving with an instance of the congruence axiom for s we obtain $\{s(f(s(o), o)) = s(s(o))\}$. Applying transitivity we have $\{f(s(o), s(o)) = s(s(o))\}$. Resolving with the negation of the conclusion we get the empty clauses.

11. The two axioms of the previous exercise define addition using the successor function, but do not suffice to prove that $0 \neq 1$ and $0 \neq 2$. Recalling what we wrote on page 58, add suitable axioms for the natural numbers and prove that $0 \neq 3$. Subsequently write two axioms for multiplication and convince yourself that the entries of the multiplication table ($1 \times 1 = 1$, $1 \times 2 = 2, \ldots, 10 \times 10 = 100$) are consequences of these axioms.

12. Prove the following statement by obtaining the empty clauses from the premise and the negation of the conclusion:

> Let R be a symmetric and transitive relation with the following property: for every x there exists y such that Rxy. Then R is reflexive.

13. Prove the following tautology by showing that it is a logical consequence of the equality axioms:

$$\forall x \forall x' \forall y \forall y' \big((x = x' \wedge y = y') \to g(f(x), y) = g(f(x'), y')\big).$$

14. Prove that the following statement is a tautology, or find a model for its negation:

$(\exists x \forall y \, Axy) \to (\forall u \exists v \, Avu).$

15. Prove that the following statement is a tautology, or find a model for its negation:

$(\forall x \forall y \ (x = a \vee y = a)) \to (\forall z \ f(a, z) = f(z, a)).$

Logical consequence, models and natural language

In each of the following exercises write in clauses the premises and the negation of the conclusion, obtaining a set S of clauses. If the conclusion is a consequence of the premises, produce a refutation of S using the equality axioms whenever needed. Otherwise find a model for S, preferably with a few elements. When the symbol $=$ occurs in S find a model with equality.

1. Every Franciscan is poor
   no accountant is poor

   ---

   there do not exist Franciscan accountants

Solution. The premise is: $\{\neg Fx, Px\}, \{\neg Ay, \neg Py\}$; the negation of the conclusion is: "there exists a Franciscan accountant", that is, "there exists an entity x that is an accountant and Franciscan", $\exists z(Az \wedge Fz)$. To eliminate the existential quantifier using Skolemisation, let's give a name, for example a for "Arturo", to one of the existing Franciscan accountants of our choice. The unique condition that we have to respect is that Arturo is a new name in this exercise.[5] So the negation of the conclusion becomes "Arturo is a Franciscan accountant", that one formalises as *two* clauses $\{Aa\}$ and $\{Fa\}$. The Herbrand universe of this exercise is the singleton set $\{a\}$. Therefore we do not have any problem in choosing an instance that quickly yields the empty clause. Instantiating the first two clauses we obtain $\{\neg Fa, Pa\}$ and $\{\neg Aa, \neg Pa\}$. We are now in propositional logic and it is easy to refute the clauses $\{\neg Fa, Pa\}, \{\neg Aa, \neg Pa\}, \{Aa\}, \{Fa\}$. Having obtained this way the empty clause we have "deduced" the conclusion of the premises using a purely formal calculus. Nobody will find a possible world simultaneously satisfying the premises and the negation of the conclusion. In fact, let $\mathcal{M} = (M, ^*)$ be a model of type $\{F, P, A, a\}$. In the possible world described by $\mathcal{M}$ the words "Franciscan", "accountant", "poor", and "Arturo" have an absolutely arbitrary meaning: F^*, P^*, A^* are just subsets of the universe M, and a^* is one of its elements. Well, even with this enormous freedom, if $\mathcal{M}$ satisfies the premises, it cannot satisfy the negation of the conclusion, and therefore $\mathcal{M}$ will satisfy the conclusion.

2. Every hare is afraid of some fox
   Bic is not afraid of anybody

   Bic is not a hare

   Hint. One formalises in $\mathcal{L}$ the premises by means of the statements $\forall x(Hx \rightarrow \exists y(Fy \wedge Axy))$ e $\forall z \neg Abz$, from which one obtains the clauses

   $$\{\neg Hx, Ff(x)\}, \quad \{\neg Hx, Axf(x)\}, \quad \{\neg Abz\}.$$

   The negation of the conclusion is the clause $\{Hb\}$. It is easy to refute these clauses: in fact, already the last three are refutable.

3. Every ungulate is a vertebrate: $\forall x(Ux \rightarrow Vx)$
   the father of every ungulate is an ungulate: $\forall y(Uy \rightarrow Up(y))$
   Bic is not a vertebrate: $\neg Vb$

   Bic is not the father of any ungulate: $\neg \exists z(Uz \wedge b = p(z))$

---

[5] The idea that *exist = have received a proper name* has a thousand-years old story that continues to be of importance in our computer: to delete a file means to remove its name.

4. For every unfortunate there is some goat that bites him
   Alf is unfortunate
   Bic is an unfortunate goat

   _____

   Bic bites Alf

   Hint. The formalisation in $\mathcal{L}$ gives the statements
   $\forall x(Ux \rightarrow \exists y(Gy \wedge Byx))$
   Ua
   $Ub \wedge Gb$

   _____

   Bba

   Then put in clauses the premises and the negation of the conclusion, and subsequently find a model that satisfies all these clauses.

5. Every A is B
   the mother of every B is B
   Ada is not A
   Ada is not B

   _____

   Ada is not the mother of any A

6. Prove the validity of the following Aristotelean syllogisms:

   a) Some A is B
      every B is C

      _____

      some C is A

   b) some A is B
      every B is C

      _____

      some A is C

   c) some A is B
      no C is B

      _____

      some A is not C

7. Every boxer fears some left-handed
   every left-handed fears some non-left-handed
   Alf is a non-left-handed boxer

   ---

   either Alf fears some non-left-handed or some non-left-handed fears Alf

8. Every progressive admires some conservative
   everybody who is not progressive is conservative
   Alf does not admire any conservative

   Alf is a conservative

   Abbreviated solution (the instances are not indicated):

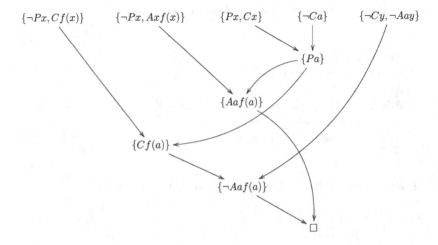

9. *Logic and barbers.* Only one among the following deductions is valid (where we introduce imaginary 'megabarbers', 'bureaucratic barbers', in short 'burobarbers', and 'megaburobarbers' who are both megabarbers and burobarbers):

   a) Every megabarber shaves all those who do not shave themselves

      ---

      there does not exist any megabarber

   b) Every burobarber shaves only those who do not shave themselves

      ---

      there does not exist any burobarber

   c) Every megaburobarber shaves all those and only those who do not shave themselves

      ---

      there does not exist any megaburobarber

Hint:

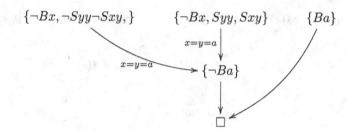

10. Every stupid listens to anybody who promises heaven and earth to him
    Alf promises heaven and earth to everybody
    _____

    there is somebody who is listened by everybody

11. Every thief is afraid of every cop
    every cop is afraid of some thief
    some cop is afraid of every thief
    Antonio is afraid of every thief and every cop
    _____

    Antonio is a cop or a thief

12. Every utopian diver is a swimmer
    no utopian dentist is a swimmer
    the mother of every utopian is utopian
    _____

    either Anne is not the mother of any utopian, or she is not a dentist, or
    she is not a diver

13. Every stupid believes anybody who talks to him
    Alf does not belief anybody
    Als is stupid
    _____

    nobody talks to Alf

14. Every A is B and every B is C
    not every C is A
    _____

    either some B is not A or some C is not B

15. Every ambitious person fights against somebody else
    Ark fights only against himself
    _____

    Ark is not ambitious

Solution:

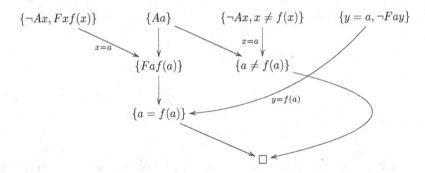

16. Every two different inhabitants of Rio Bo have different mothers
    Aldo and Mr Palazzi have the same mother

    _____

    either Aldo or Mr Palazzi is not an inhabitant of Rio Bo

17. Every rigorous person admires some mathematician
    every mathematician admires some rigorous person
    Alf admires only himself

    _____

    either Alf is not a mathematician or is a rigorous person

    Abbreviated solution, without indication of the ground instances, and without explicit use of the congruence clauses:

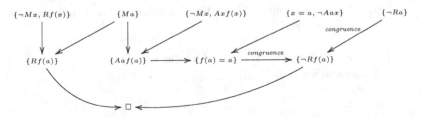

18. Every boxer is afraid of some left-handed
    every left-handed is afraid of some non-left-handed
    Alf is a left-handed boxer

    _____

    Alf is afraid of at least two persons

19. There are precisely two elements in A
    R is a reflexive and transitive relation over A

    _____

    R is symmetric

20. Write the three axioms (reflexivity, antisymmetry and transitivity) that define a partial order relation M. Obtain using the refutational method some elementary properties of M. The simplest ones are: there exists at most one largest element; if there exists a largest element, then it is also maximal, that is, it is not strictly dominated by any element.

21. With reference to Exercise 5 on page 51, compare the following two deductions:

Martians do not exist: $\neg \exists x M_1 x$

---

it is not true that if Alf travels he meets a Martian:
$\neg(Ta \to \exists y(M_1 y \wedge M_2 ay))$

Martians do not exist

---

whether he travels or not, Alf does not meet Martians

Prove that precisely one of these deductions is valid.

22. Consider the phrase:

> *There is a person such that if he votes for Octavio, then everybody votes for Octavio.*

Write in clauses its negation and refute it. So this phrase is a tautology, even though our daily way of understanding "if" and "there is somebody" does not help us in understanding that this phrase is infallibly true in all possible worlds. Can this tautology be used by Octavio to get all the votes?

23. For each of the following phrases, if it is not a tautology, then find a model that satisfies its negation; if it is a tautology, then refute its negation:

   (a) there is a person such that if he votes for Octavio, then somebody else votes for Octavio;

   (b) there is a person such that if he votes for Octavio, then everybody votes for Octavio, but if he does not vote for Octavio, then nobody votes for Octavio;

   (c) there is a person such that if he votes for Octavio, then everybody votes for Octavio, and there is a person such that if he does not vote for Octavio, then nobody votes for Octavio;

(d) there is a person such that if he votes for Octavio, then Octavio votes for himself;

(e) there is a person such that if he does not vote for Octavio, then Octavio votes for himself;

(f) there is a person such that if he does not vote for Octavio, then Octavio does not vote for himself.

17

Final Remarks

We have prepared a formidable symbolic apparatus, with its logical calculus, and we can now launch it in the vast field of mathematics for which it was constructed. For example, if we wish to dedicate ourselves to the study of the problem of twin prime numbers $p, p + 2$ introduced on page 57, we cannot do anything else than accept the axioms for natural numbers, or for sets, and subsequently get down to calculate the consequences of the axioms – mentally, or with the help of lemmas and theorems previously obtained, or even with the help of a computer that generates for us pairs of twin primes with more than hundred thousand digits, hence suggesting that there are infinitely many of such pairs. The completeness theorem assures us that no consequence of the axioms will escape the logical calculus.

The axiomatic method is for mathematicians what the experimental method is for physicists. It has proved to be so powerful that it has also affected physics. For example, the analysis of the possible geometries obtained by dropping one of the Euclid axioms produced seemingly esoteric models which, however, have found use in physics later on. In 1899 Hilbert published a book (*Grundlagen der Geometrie*) dedicated to the axiomatic foundation of geometry. In 1903 Poincaré commented:

> This was not so easy as one might suppose; there are the axioms which one sees and those which one does not see, which are introduced unconsciously and without being noticed. [...] Is the list of Professor Hilbert final? We may take it to be so, for it seems to have been drawn up with care.

Poincaré seems to be asking himself whether the list of axioms of Hilbert is complete, that is, whether they suffice to decide all geometric conjectures. Perhaps Poincaré even asks himself whether this question is important for the development of geometry. The point is that if the list is not complete, somebody will have to make an addition – and surely the axioms to be added will be surprising since they escaped even Professor Hilbert.

Mundici D.: Logic: a Brief Course.
DOI 10.1007/978-88-470-2361-1_17, © Springer-Verlag Italia 2012

The completeness problem also arises for the numbers: starting from Descartes the real numbers, with their equations and disequations were shown to be crucial for treating every geometric entity, to the point of becoming indispensable for defining geometric entities in spaces of arbitrary dimensions. Since the real numbers are constructed from the rational numbers and these are constructed from the integers, a spontaneous question arose: *what are the natural numbers?* And so Dedekind, Peano and others proposed axioms for natural numbers even before the *Grundlagen der Geometrie*.

With a handful of simple axioms for addition and multiplication, like those on page 58, the logical calculus allows us to recover the hundred mini-theorems of the multiplication table. Continuing this way we will be able to prove several more profound results, including the Euclid theorem about the infinity of the prime numbers. But our axioms will be in any case a caricature of the Minimum Principle, equivalent to the Induction Principle, stating that *"every nonempty set X of natural numbers has a least element."* This universal quantifier $\forall X$ is very different from the quantifier $\forall x$ used until now, that makes the variable x vary over the universe M of the model and not over the subsets of M. The necessity to express the Induction Principle using the universal quantifier of the logic $\mathcal{L}$ requires that we enlarge the horizon, introducing axioms not anymore for the totality of natural numbers but for the totality of sets, and defining $\mathbb{N}$ and its subsets as the *elements* of this totality.

When preparing a list $\mathcal{A}$ of axioms for the universe of all sets, or for any system of mathematical entities, one has to absolutely avoid the:

Major Evil: Incoherence (= Refutability). Incoherence means that the logical calculus can derive from $\mathcal{A}$ the empty clause. As a result for every statement E, both E and $\neg E$ are consequences of $\mathcal{A}$. As $\mathcal{A}$ is unsatisfiable, it does not distinguish true from false and does not axiomatise any model.

For example, Frege, the father of predicate logic, conjectured that for every formula $F(x)$ of $\mathcal{L}$ there exists a set consisting precisely of the elements x that satisfy $F(x)$. In particular, when $F(x)$ is the formula $x \notin x$, that expresses the property of not being an element of itself, we will have to admit the axiom $\exists s \, \forall x (x \in s \leftrightarrow x \notin x)$. But the logical calculus easily derives from this statement the empty clause. (*The final exercise of this course.*)

Having avoided the major evil, one should avoid the other:

Minor Evil, Incompleteness. Incompleteness means that there exists a proposition E that is *undecidable* in $\mathcal{A}$, in the sense that neither E nor $\neg E$ is a consequence of $\mathcal{A}$. Such a set of axioms $\mathcal{A}$ is said to be *incomplete* because it does not succeed to decide which among E and $\neg E$ holds in the model that we were hoping to axiomatise.

In 1931 Gödel proved the admirable *Incompleteness Theorem*, for the full understanding of which a second course of logic is necessary, in which one defines mathematically the notions of "algorithm" and of "formal system". A corollary expressible using the means that are at our disposal states that each of our attempts to axiomatise sets or natural numbers that succeeds to avoid the major evil, will not avoid the minor evil.

More precisely, *let $\mathcal{A}$ be a finite set of statements of a type $\tau \supseteq \{f, g\}$, and let $\mathcal{N} = (\mathbb{N}, *)$ be a model of natural numbers with $f^* = $ addition and $g^* = $ multiplication. If $\mathcal{N} \models \mathcal{A}$, then there exists a statement E of type τ such that $\mathcal{N} \models E$ but E is not a consequence of $\mathcal{A}$.*

So, although E is a true arithmetic statement, it is not a consequence of $\mathcal{A}$. But neither $\neg E$ is a consequence of $\mathcal{A}$, because otherwise the assumption $\mathcal{N} \models \mathcal{A}$ would imply $\mathcal{N} \models \neg E$, and therefore $\mathcal{N} \models E \wedge \neg E$, which is impossible. In conclusion, $\mathcal{N}$ does not admit a finite and complete list of axioms.

Gödel Incompleteness Theorem, of which this corollary is only a foretaste, is one of the most important results of twentieth century mathematics. It yields a distinction between numbers viewed as a means to count, as on an abacus, and numbers viewed as "numerals", that is, as the structure $\mathcal{N}$ that is at the base of our perceptions and manipulations. This distinction perhaps goes back to the Pythagorean school, and through Plotinus was resumed by Saint Augustine (Confessions, §10.12) and others.

The logical calculus $\mathcal{L}$ is certainly able to derive *all* the consequences of the set $\mathcal{A}$ of our axioms, but $\mathcal{A}$ is deficient: it leaves open some problems concerning addition and multiplication – while we perhaps have fooled ourselves that our axioms would capture the whole truth about the four arithmetic operations. Which property of addition and multiplication, expressible in $\mathcal{L}$, could have eluded the mathematicians for all these millennia of reflection on $\mathcal{N}$? It is hard to say, especially since Gödel Incompleteness Theorem tells us that if $\mathcal{A}$ were substituted by a richer set $\mathcal{A}'$ of axioms, also $\mathcal{A}'$ would leave open some conjecture E', provided $\mathcal{A}'$ is coherent.

We might then think that incompleteness is due to the expressive poverty of $\mathcal{L}$, which is clearly shown by the final corollaries of this course. Why to stop at predicate logic with equality $\mathcal{L}$? One reason could be the theorem proved in 1965 by Lindström, according to which *every extension of the logic $\mathcal{L}$ either does not have the compactness property or does not satisfy the Löwenheim property*, that we proved for $\mathcal{L}$ in Corollary 16.13. Therefore, every enrichment of $\mathcal{L}$ will have to take into account the loss of at least one of these two important ingredients of the completeness of the logical calculus $\mathcal{L}$. For example, adding a new quantifier $\exists_\infty x$ that states "there are infinitely many x", we could write $\forall n \, \neg \exists_\infty x \; x < n$, and therefore ban the nonstandard numerals, avoiding the pathology of Corollary 16.14. But as we have repeatedly seen in this course, to each enrichment of the symbolic apparatus there corresponds an increased complication of the logical calculus: the calculus for the connectives $\vee, \wedge, \neg$ and the propositional variables is considerably simpler than the calculus operating on a richer symbolic apparatus of $\vee, \wedge, \neg, \forall, \exists$ with predicates, function, variables and constants. A further complication comes from incorporating the equality symbol, to reach the full expressive power of $\mathcal{L}$.

The price to pay for adding to $\mathcal{L}$ the quantifier $\exists_\infty$ is too high: for the so obtained logic there cannot exist a complete logical calculus. Also, if we tried other, more expressive logics and other axioms, Gödel Incompleteness Theorem tells us that the truth contained in the additive-multiplicative totality of the natural numbers, however axiomatised, is not *fully* accessible by any logical calculus. What holds for the structure $(\mathbb{N}, +, \cdot, 0, 1)$ obviously applies to every more complex structure.

But perhaps, not even a mathematical truth as profound and general as this one touches upon the problems posed by the protagonist of Exercise 8 on page 87.

Index

Collana Unitext – La Matematica per il 3+2

Series Editors:
A. Quarteroni (Editor-in-Chief)
L. Ambrosio
P. Biscari
C. Ciliberto
G. van der Geer
G. Rinaldi
W.J. Runggaldier

Editor at Springer:
F. Bonadei
francesca.bonadei@springer.com

As of 2004, the books published in the series have been given a volume number. Titles in grey indicate editions out of print.
As of 2011, the series also publishes books in English.

A. Bernasconi, B. Codenotti
Introduzione alla complessità computazionale
1998, X+260 pp, ISBN 88-470-0020-3

A. Bernasconi, B. Codenotti, G. Resta
Metodi matematici in complessità computazionale
1999, X+364 pp, ISBN 88-470-0060-2

E. Salinelli, F. Tomarelli
Modelli dinamici discreti
2002, XII+354 pp, ISBN 88-470-0187-0

S. Bosch
Algebra
2003, VIII+380 pp, ISBN 88-470-0221-4

S. Graffi, M. Degli Esposti
Fisica matematica discreta
2003, X+248 pp, ISBN 88-470-0212-5

S. Margarita, E. Salinelli
MultiMath - Matematica Multimediale per l'Università
2004, XX+270 pp, ISBN 88-470-0228-1

A. Quarteroni, R. Sacco, F.Saleri
Matematica numerica (2a Ed.)
2000, XIV+448 pp, ISBN 88-470-0077-7
2002, 2004 ristampa riveduta e corretta
(1a edizione 1998, ISBN 88-470-0010-6)

13. A. Quarteroni, F. Saleri
    Introduzione al Calcolo Scientifico (2a Ed.)
    2004, X+262 pp, ISBN 88-470-0256-7
    (1a edizione 2002, ISBN 88-470-0149-8)

14. S. Salsa
    Equazioni a derivate parziali - Metodi, modelli e applicazioni
    2004, XII+426 pp, ISBN 88-470-0259-1

15. G. Riccardi
    Calcolo differenziale ed integrale
    2004, XII+314 pp, ISBN 88-470-0285-0

16. M. Impedovo
    Matematica generale con il calcolatore
    2005, X+526 pp, ISBN 88-470-0258-3

17. L. Formaggia, F. Saleri, A. Veneziani
    Applicazioni ed esercizi di modellistica numerica
    per problemi differenziali
    2005, VIII+396 pp, ISBN 88-470-0257-5

18. S. Salsa, G. Verzini
    Equazioni a derivate parziali – Complementi ed esercizi
    2005, VIII+406 pp, ISBN 88-470-0260-5
    2007, ristampa con modifiche

19. C. Canuto, A. Tabacco
    Analisi Matematica I (2a Ed.)
    2005, XII+448 pp, ISBN 88-470-0337-7
    (1a edizione, 2003, XII+376 pp, ISBN 88-470-0220-6)

20. F. Biagini, M. Campanino
    Elementi di Probabilità e Statistica
    2006, XII+236 pp, ISBN 88-470-0330-X

21. S. Leonesi, C. Toffalori
    Numeri e Crittografia
    2006, VIII+178 pp, ISBN 88-470-0331-8

22. A. Quarteroni, F. Saleri
    Introduzione al Calcolo Scientifico (3a Ed.)
    2006, X+306 pp, ISBN 88-470-0480-2

23. S. Leonesi, C. Toffalori
    Un invito all'Algebra
    2006, XVII+432 pp, ISBN 88-470-0313-X

24. W.M. Baldoni, C. Ciliberto, G.M. Piacentini Cattaneo
    Aritmetica, Crittografia e Codici
    2006, XVI+518 pp, ISBN 88-470-0455-1

25. A. Quarteroni
    Modellistica numerica per problemi differenziali (3a Ed.)
    2006, XIV+452 pp, ISBN 88-470-0493-4
    (1a edizione 2000, ISBN 88-470-0108-0)
    (2a edizione 2003, ISBN 88-470-0203-6)

26. M. Abate, F. Tovena
    Curve e superfici
    2006, XIV+394 pp, ISBN 88-470-0535-3

27. L. Giuzzi
    Codici correttori
    2006, XVI+402 pp, ISBN 88-470-0539-6

28. L. Robbiano
    Algebra lineare
    2007, XVI+210 pp, ISBN 88-470-0446-2

29. E. Rosazza Gianin, C. Sgarra
    Esercizi di finanza matematica
    2007, X+184 pp, ISBN 978-88-470-0610-2

30. A. Machì
Gruppi – Una introduzione a idee e metodi della Teoria dei Gruppi
2007, XII+350 pp, ISBN 978-88-470-0622-5
2010, ristampa con modifiche

31 Y. Biollay, A. Chaabouni, J. Stubbe
Matematica si parte!
A cura di A. Quarteroni
2007, XII+196 pp, ISBN 978-88-470-0675-1

32. M. Manetti
Topologia
2008, XII+298 pp, ISBN 978-88-470-0756-7

33. A. Pascucci
Calcolo stocastico per la finanza
2008, XVI+518 pp, ISBN 978-88-470-0600-3

34. A. Quarteroni, R. Sacco, F. Saleri
Matematica numerica (3a Ed.)
2008, XVI+510 pp, ISBN 978-88-470-0782-6

35. P. Cannarsa, T. D'Aprile
Introduzione alla teoria della misura e all'analisi funzionale
2008, XII+268 pp, ISBN 978-88-470-0701-7

36. A. Quarteroni, F. Saleri
Calcolo scientifico (4a Ed.)
2008, XIV+358 pp, ISBN 978-88-470-0837-3

37. C. Canuto, A. Tabacco
Analisi Matematica I (3a Ed.)
2008, XIV+452 pp, ISBN 978-88-470-0871-3

38. S. Gabelli
Teoria delle Equazioni e Teoria di Galois
2008, XVI+410 pp, ISBN 978-88-470-0618-8

39. A. Quarteroni
Modellistica numerica per problemi differenziali (4a Ed.)
2008, XVI+560 pp, ISBN 978-88-470-0841-0

40. C. Canuto, A. Tabacco
Analisi Matematica II
2008, XVI+536 pp, ISBN 978-88-470-0873-1
2010, ristampa con modifiche

41. E. Salinelli, F. Tomarelli
Modelli Dinamici Discreti (2a Ed.)
2009, XIV+382 pp, ISBN 978-88-470-1075-8

42. S. Salsa, F.M.G. Vegni, A. Zaretti, P. Zunino
Invito alle equazioni a derivate parziali
2009, XIV+440 pp, ISBN 978-88-470-1179-3

43. S. Dulli, S. Furini, E. Peron
Data mining
2009, XIV+178 pp, ISBN 978-88-470-1162-5

44. A. Pascucci, W.J. Runggaldier
Finanza Matematica
2009, X+264 pp, ISBN 978-88-470-1441-1

45. S. Salsa
Equazioni a derivate parziali – Metodi, modelli e applicazioni (2a Ed.)
2010, XVI+614 pp, ISBN 978-88-470-1645-3

46. C. D'Angelo, A. Quarteroni
Matematica Numerica – Esercizi, Laboratori e Progetti
2010, VIII+374 pp, ISBN 978-88-470-1639-2

47. V. Moretti
Teoria Spettrale e Meccanica Quantistica – Operatori in spazi di Hilbert
2010, XVI+704 pp, ISBN 978-88-470-1610-1

48. C. Parenti, A. Parmeggiani
Algebra lineare ed equazioni differenziali ordinarie
2010, VIII+208 pp, ISBN 978-88-470-1787-0

49. B. Korte, J. Vygen
Ottimizzazione Combinatoria. Teoria e Algoritmi
2010, XVI+662 pp, ISBN 978-88-470-1522-7

50. D. Mundici
Logica: Metodo Breve
2011, XII+126 pp, ISBN 978-88-470-1883-9

51. E. Fortuna, R. Frigerio, R. Pardini
Geometria proiettiva. Problemi risolti e richiami di teoria
2011, VIII+274 pp, ISBN 978-88-470-1746-7

The online version of the books published in this series is available at SpringerLink.
For further information, please visit the following link:
http://www.springer.com/series/5418